19/20

SLC

DSP Applications Using C
and the TMS320C6x DSK

TOPICS IN DIGITAL SIGNAL PROCESSING

DSP Applications Using C and the TMS320C6x DSK

Rulph Chassaing

A Wiley–Interscience Publication
JOHN WILEY & SONS, INC.

Copyright © 2002 by John Wiley & Sons, Inc., New York. All rights reserved.

Published simultaneously in Canada.

For ordering and customer service, call 1-800-CALL-WILEY.

Library of Congress Cataloging-in-Publication Data:
Chassaing, Rulph.
 DSP applications using C and the TMS320C6x DSK / Rulph Chassaing.
  p. cm. – (Topics in digital signal processing)
 Includes bibliographical references and index.
 ISBN 0-471-20754-3 (cloth : alk. paper)
  1. Signal processing—Digital techniques. 2. Texas Instruments TMS320 series microprocessors. 3. C (Computer program language) I. Title. II. Series.
TK5102.9 .C47423 2002
621.382′2–dc21   2001046889

Printed in the United States of America.

10 9 8 7 6 5 4 3 2 1

Contents

Preface

Digital signal processors, such as the TMS320 family of processors, are used in a wide range of applications, such as in communications, controls, speech processing, and so on. They are used in fax transmission, modems, cellular phones, and other devices. These devices have also found their way into the university classroom, where they provide an economical way to introduce real-time digital signal processing (DSP) to the student.

Texas Instruments recently introduced the TM320C6x processor, based on the very-long-instruction-word (VLIW) architecture. This newer architecture supports features that facilitate the development of efficient high-level language compilers. Throughout the book we refer to the C/C++ language simply as C. Although TMS320C6x/assembly language can produce fast code, problems with documentation and maintenance may exist. With the available C compiler, the programmer must consider to "let the tools do the work." After that, if the programmer is not satisfied, Chapters 3 and 8 and the last few examples in Chapter 4 can be very useful.

This book is intended primarily for senior undergraduate and first-year graduate students in electrical and computer engineering and as a tutorial for the practicing engineer. It is written with the conviction that the principles of DSP can best be learned through interaction in a laboratory setting, where students can appreciate the concepts of DSP through real-time implementation of experiments and projects. The background assumed is a course in linear systems and some knowledge of C.

Most chapters begin with a theoretical discussion, followed by representative examples that provide the necessary background to perform the concluding experiments. There are a total of 76 solved programming examples, most using C code, with a few in assembly and linear assembly code. A list of these examples appears on page xv. Several sample projects are also discussed.

Programming examples are included throughout the text. This can be useful to the reader who is familiar with both DSP and C programming but who is not necessarily an expert in both.

This book can be used in the following ways:

1. For a DSP course with a laboratory component, using Chapters 1 to 7 and Appendices D to F. If needed, the book can be supplemented with some additional theoretical materials, since the book's emphasis is on the practical aspects of DSP. It is possible to cover Chapter 7 on adaptive filtering, following Chapter 4 on FIR filtering (since there is only one example in Chapter 7 that uses material from Chapter 5). It is my conviction that adaptive filtering (Chapter 7) should be incorporated into an undergraduate course in DSP.

2. For a laboratory course using many of the examples and experiments from Chapters 1 to 7. The beginning of the semester can be devoted to short programming examples and experiments and the remainder of the semester used for a final project.

3. For a senior undergraduate or first-year graduate design project course, using Chapters 1 to 5, selected materials from Chapters 6 to 9, and Appendices D to F.

4. For the practicing engineer as a tutorial, and for workshops and seminars, using selected materials throughout the book.

In Chapter 1 we introduce the tools through three programming examples. These tools include the powerful Code Composer Studio (CCS) provided with the TMS320C6711 DSP starter kit (DSK). It is essential to perform these three examples before proceeding to subsequent chapters. They illustrate the capabilities of CCS for debugging, plotting in both the time and frequency domains, and other matters.

In Chapter 2 we illustrate input and output (I/O) with the codec on the DSK board through many programming examples. Alternative I/O with a stereo audio codec that interfaces with the DSK is described. Chapter 3 covers the architecture and the instructions available for the TMS320C6x processor. Special instructions and assembler directives that are useful in DSP are discussed. Programming examples using both assembly and linear assembly are included in this chapter.

In Chapter 4 we introduce the z-transform and discuss finite impulse response (FIR) filters and the effect of window functions on these filters. Chapter 5 covers infinite impulse response (IIR) filters. Programming examples to implement real-time FIR and IIR filters are included.

Chapter 6 covers the development of the fast Fourier transform (FFT). Programming examples on FFT are included. In Chapter 7 we demonstrate the usefulness of the adaptive filter for a number of applications with least mean squares (LMS). Programming examples are included to illustrate the gradual cancellation of noise or system identification. Chapter 8 illustrates techniques for code optimization. In Chapter 9 we discuss a number of DSP applications and student projects.

A disk included with this book contains all the programs discussed. See page xix for a list of the folders that contain the support files for all the examples.

Over the last six years, faculty members from over 150 institutions have taken my "DSP and Applications" workshops. These workshops were supported for three years by grants from the National Science Foundation (NSF) and subsequently, by Texas Instruments. I am thankful to NSF, Texas Instruments, and the participating faculty members for their encouragement and feedback. I am grateful to Dr. Donald Reay of Heriot-Watt University, who contributed several examples during his review of the book. I appreciate the many suggestions made by Dr. Robert Kubichek of the University of Wyoming during his review of the book. I also thank Dr. Darrell Horning of the University of New Haven, with whom I coauthored the text *Digital Signal Processing with the TMS320C25*, for introducing me to "book writing." I thank all the students (at Roger Williams University, University of Massachusetts, Dartmouth, and Worcester Polytechnic Institute) who have taken my real-time DSP and senior design project courses, based on the TMS320 processors, over the last 16 years. I am particularly indebted to two former students, Bill Bitler and Peter Martin, who have worked with me over the years. The laboratory assistance of Walter J. Gomes III in several workshops and during the development of many examples has been invaluable. The continued support of many people from Texas Instruments is also very much appreciated: Maria Ho and Christina Peterson, in particular, have been very supportive of this book. I would be remiss if I did not mention the librarians in Herkimer, New York (where I was stranded for two weeks) for the use of their facility to write Chapter 8.

RULPH CHASSAING
Chassaing@msn.com

List of Examples

Programs/Files on Accompanying Disk

A list of the folders included on the accompanying disk is shown below. The folders contain the programs/files for all the examples/projects covered in the book.

Exploring - myprojects

File   Edit   View   Go   Favorites   Tools   Help

Back   Forward   Up   Cut   Copy   Paste   Undo   Delete   Properties   Views

Address C:\ti\myprojects

Folders

- ti
  - bin
  - c6000
  - cc
  - docs
  - drivers
  - examples
  - myprojects
    - Adaptc
    - AdaptIDFIR
    - adaptidFIRw
    - AdaptIDIIR
    - Adaptnoise
    - Adaptnoise_pcm
    - Adaptpredict
    - Adaptpredict_pcm
    - Aliasing
    - Am
    - Dft
    - Dotp4

Adaptc	AdaptIDFIR	adaptidFIRw	AdaptIDIIR
Adaptnoise	Adaptnoise_pcm	Adaptpredict	Adaptpredict_pcm
Aliasing	Am	Dft	Dotp4
dotp4a	dotp4clasm	Dotpintrinsic	Dotpipedfix
Dotpipedfloat	Dotpnp	Dotpnpfloat	dotpopt
Dotpp	Dotppfloat	echo	Echo_control
Factclasm	Factorial	Fastconvo	Fastconvosim
FFT256c	FFTsinetable	Fir	FIR_pcm
Fir3lp	FIR4types	FIR4ways	FIRbuf
Fircasm	FIRcasmfast	FIRcirc	FIRcirc_ext
FIRinverse	Firprn	FIRPRNbuf	GraphicEQ
Iir	IIRinverse	loop_intr	loop_intr_pcm
loop_poll	loop_poll_pcm	loop_print	Loop_store
Miscellaneous	Noise_gen	Notch2	Pll
Ramp	Ramptable	Scram16k	sin1500MATL
sine2sliders	sine4_poll	sine8_buf	sine8_intr
Sinegen_table	SinegenDE	Squarewave	Sum
Support	sweep8000	SweepDE	two_tones
Twosum	Twosumfix	Twosumfloat	Twosumlasmfix
Twosumlasmfloat	readme		

DSP Applications Using C
and the TMS320C6x DSK

1

DSP Development System

- Testing the software and hardware tools with Code Composer Studio
- Use of the TMS320C6711 DSK
- Programming examples to test the tools

Chapter 1 introduces several tools available for digital signal processing (DSP). These tools include the popular Code Composer Studio (CCS), which provides an integrated development environment (IDE); the DSP starter kit (DSK) with the TMS320C6711 floating-point processor onboard and complete support for input and output. Three examples are included to test both the software and hardware tools included with the DSK.

1.1 INTRODUCTION

Digital signal processors such as the TMS320C6x (C6x) family of processors are like fast special-purpose microprocessors with a specialized type of architecture and instruction set appropriate for signal processing. The C6x notation is used to designate a member of Texas Instruments' (TI) TMS320C6000 family of digital signal processors. The architecture of the C6x digital signal processor is very well suited for numerically intensive calculations. Based on a very-long-instruction-word (VLIW) architecture, the C6x is considered to be TI's most powerful processor.

Digital signal processors are used for a wide range of applications, from communications and controls to speech and image processing. They are found in cellular phones, fax/modems, disk drives, radio, and so on. These processors have become the product of choice for a number of consumer applications, since they have become very cost-effective. They can handle different tasks, since they can be

1

reprogrammed readily for a different application. DSP techniques have been very successful because of the development of low-cost software and hardware support. For example, modems and speech recognition can be less expensive using DSP techniques.

DSP processors are concerned primarily with real-time signal processing. Real-time processing means that the processing must keep pace with some external event; whereas non-real-time processing has no such timing constraint. The external event to keep pace with is usually the analog input. While analog-based systems with discrete electronic components such as resistors can be more sensitive to temperature changes, DSP-based systems are less affected by environmental conditions such as temperature. DSP processors enjoy the advantages of microprocessors. They are easy to use, flexible, and economical.

A number of books and articles have been published that address the importance of digital signal processors for a number of applications [1–20]. Various technologies have been used for real-time processing, from fiber optics for very high frequency to DSP processors very suitable for the audio-frequency range. Common applications using these processors have been for frequencies from 0 to 20kHz. Speech can be sampled at 8kHz (how quickly samples are acquired), which implies that each value sampled is acquired at a rate of 1/(8kHz) or 0.125ms. A commonly used sample rate of a compact disk is 44.1 kIIz. A/D-based boards in the megahertz sampling rate range are currently available.

The basic system consists of an analog-to-digital converter (ADC) to capture an input signal. The resulting digital representation of the captured signal is then processed by a digital signal processor such as the C6x and then output through a digital-to-analog converter (DAC). Also included within the basic system is a special input filter for antialiasing to eliminate erroneous signals, and an output filter to smooth or reconstruct the processed output signal.

1.2 DSK SUPPORT TOOLS

Most of the work presented in this book involves the design of a program to implement a DSP application. To perform the experiments, the following tools are used:

1. *TI's DSP starter kit (DSK)*. The DSK package includes:
   (a) Code Composer Studio (CCS), which provides the necessary software support tools. CCS provides an integrated development environment (IDE), bringing together the C compiler, assembler, linker, debugger, and so on.
   (b) A board, shown in Figure 1.1a, that contains the TMS320C6711 (C6711) floating-point digital signal processor as well as a 16-bit codec for input and output (I/O) support.
   (c) A parallel cable (DB25) that connects the DSK board to a PC.
   (d) A power supply for the DSK board.

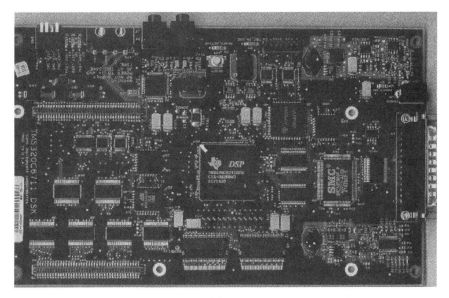

(a)

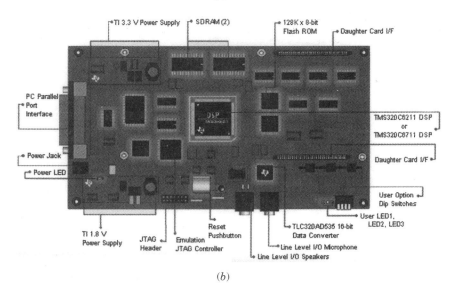

(b)

FIGURE 1.1. TMS320C6711-based DSK board: (a) board; (b) diagram (Courtesy of Texas Instruments).

2. *An IBM-compatible PC.* The DSK board connects to the parallel port of the PC through the DB25 cable included with the DSK package.

3. *An oscilloscope, signal generator, and speakers.* A signal/spectrum analyzer is optional. Shareware utilities are available that utilize the PC and a sound card to create a virtual instrument such as an oscilloscope, a function generator, or a spectrum analyzer.

All the files/programs listed and discussed in this book (except the student project files in Chapter 9) are included on the accompanying disk. Most of the examples can also run on the fixed-point C6211-based DSK (which has been discontinued). A list of all the examples is given on pages xv–xviii.

1.2.1 DSK Board

The DSK package is powerful, yet relatively inexpensive ($295), with the necessary hardware and software support tools for real-time signal processing [21–33]. It is a complete DSP system. The DSK board, with an approximate dimension of 5 × 8 inches, includes the C6711 floating-point digital signal processor [22] and a 16-bit codec AD535 for input and output.

The onboard codec AD535 [34] uses a sigma–delta technology that provides analog-to-digital conversion (ADC) and digital-to-analog conversion (DAC). A 4-MHz clock onboard the DSK connects to this codec to provide a fixed sampling rate of 8 kHz.

A daughter card expansion is also provided on the DSK board. We will illustrate input and output by plugging an audio daughter card based on the PCM3003 stereo codec (not included with the DSK package) into an 80-pin connector on the DSK board. The audio daughter card is available from Texas Instruments and is described in Appendix F. The PCM3003 codec has variable sample rates up to 72 kHz and can be useful for applications requiring higher sampling rates and two accessible input and output channels.

The DSK board includes 16 MB (megabytes) of synchronous dynamic RAM (SDRAM) and 128 kB (kilobytes) of flash ROM. Two connectors on the board provide input and output and are labeled IN (J7) and OUT (J6), respectively. Three of the four user dip switches on the DSK board can be read from a program (a project example on voice scrambling makes use of these switches). The onboard clock is 150 MHz. Also onboard the DSK are voltage regulators that provide 1.8 V for the C6711 core and 3.3 V for its memory and peripherals.

1.2.2 TMS320C6711 Digital Signal Processor

The TMS320C6711 (C6711) is based on the very-long-instruction-word (VLIW) architecture, which is very well suited for numerically intensive algorithms. The internal program memory is structured so that a total of eight instructions can be fetched every cycle. For example, with a clock rate of 150 MHz, the C6711 is capable of fetching eight 32-bit instructions every 1/(150 MHz) or 6.66 ns.

Features of the C6711 include 72 kB of internal memory, eight functional or execution units composed of six ALUs and two multiplier units, a 32-bit address bus to address 4 GB (gigabytes), and two sets of 32-bit general-purpose registers.

The C67xx (such as the C6701 and C6711) belong to the family of the C6x floating-point processors; whereas the C62xx and C64xx belong to the family of the C6x fixed-point processors. The C6711 is capable of both fixed- and floating-point processing. The architecture and instruction set of the C6711 are discussed in Chapter 3.

1.3 CODE COMPOSER STUDIO

The Code Composer Studio (CCS) provides an integrated development environment (IDE) to incorporate the software tools. CCS includes tools for code generation, such as a C compiler, an assembler, and a linker. It has graphical capabilities and supports real-time debugging. It provides an easy-to-use software tool to build and debug programs.

The C compiler compiles a C source program with extension . c to produce an assembly source file with extension . asm. The assembler assembles an . asm source file to produce a machine language object file with extension . obj. The linker combines object files and object libraries as input to produce an executable file with extension . out. This executable file represents a linked common object file format (COFF), popular in Unix-based systems and adopted by several makers of digital signal processors [21]. This executable file can be loaded and run directly on the C6711 processor.

To create an application project, one can "add" the appropriate files to the project. Compiler/linker options can readily be specified. A number of debugging features are available, including setting breakpoints and watching variables, viewing memory, registers, and mixed C and assembly code, graphing results, and monitoring execution time. One can step through a program in different ways (step into, or over, or out).

Real-time analysis can be performed using real-time data exchange (RTDX) associated with DSP/BIOS (Appendix G). RTDX allows for data exchange between the host and the target and analysis in real time without stopping the target. Key statistics and performance can be monitored in real time. Through the Joint Team Action Group (JTAG), communication with on-chip emulation support occurs to control and monitor program execution. The C6711 DSK board includes a JTAG emulator interface.

1.3.1 CCS Installation and Support

Use the parallel (printer) cable DB25 to connect the DSK board (J2) to the parallel port on the PC, such as LPT1 or LPT2. Use the 5-V adapter included with the DSK package to connect to the power connector J4, to turn on the DSK. Install CCS with the CD-ROM included with the DSK, preferably using the c:\ti structure (as default).

The CCS icon should be on the desktop as "CCS 2 ['C 6000]" and is used to launch CCS. The code generation tools (C compiler, assembler, linker) Version 4.1 are used.

On power, the three LEDs located near the four user dip switches should count from 1 to 7 (binary).

CCS provides useful documentations included with the DSK package on the following (see the Help icon):

1. Code generation tools (compiler, assembler, linker, etc.)
2. Tutorials on CCS, compiler, RTDX, advanced DSP/BIOS
3. DSP instructions and registers
4. Tools on RTDX, DSP/BIOS, and so on.

An extensive amount of support material (*pdf* files) is included with CCS (see Refs. 22 to 34). There are also a few examples included with CCS, such as a confidence test example for the DSK, an audio example, and an example associated with the onboard flash.

CCS Version 2 was used to build and test the examples included in this book. A number of files included in the following subfolders/directories within *c:\ti* can be very useful:

1. *docs*: contains documentation and manuals.
2. *myprojects*: supplied for your projects. All the programs and projects discussed in this book can be placed within this subdirectory.
3. *c6000\cgtools*: contains code generation tools.
4. *bin*: contains many utilities.
5. *c6000\examples*: contains examples included with CCS.
6. *c6000\RTDX*: contains support files for real-time data transfer.
7. *c6000\bios*: contains support files for DSP/BIOS.

1.3.2 Useful Types of Files

You will be working with a number of files with different extensions. They include:

1. file.pjt: to create and build a project named file.
2. file.c: C source program.
3. file.asm: assembly source program created by the user, by the C compiler, or by the linear optimizer.
4. file.sa: linear assembly source program. The linear optimizer uses *file.sa* as input to produce an assembly program *file.asm*.
5. file.h: header support file.
6. file.lib: library file, such as the run-time support library file rts6701.lib.
7. file.cmd: linker command file that maps sections to memory.
8. file.obj: object file created by the assembler.

9. file.out: executable file created by the linker to be loaded and run on the processor.

1.4 PROGRAMMING EXAMPLES TO TEST THE DSK TOOLS

Three programming examples are introduced to illustrate some of the features of CCS and the DSK board. The primary focus is to become familiar with both the software and hardware tools. It is strongly suggested that you complete these three examples before proceeding to subsequent chapters.

1.4.1 Quick Test of DSK

Launch CCS from the icon on the desktop. Press GEL → Check DSK → Quick Test. The Quick Test can be used for confirmation of correct operation and installation. The following message is then displayed:

Switches: 7
Revision: 2
Target is OK

This assumes that the first three switches, USER_SW1, USER_SW2, and USER_SW3, are all in the up (ON) position. Change the switches to $(1\,1\,0\,x)_2$ so that the first two switches are up (press the third switch down). The fourth switch is not used.

Repeat the procedure to select GEL → Check DSK → Quick Test and verify that the value of the switches is now 3 (with the display "Switches: 3"). You can set the value of the first three user switches from 0 to 7. Within your program you can then direct the execution of your code based on these eight values. Note that the Quick Test cycles the LEDs three times.

A confidence test program example is included with the DSK to test and verify proper operation of the major components of the DSK, such as interrupts, LEDs, SDRAM, DMA, serial ports, and timers.

Alternative Quick Test of DSK

1. Open/launch CCS from the icon on the desktop. Select File → Load Program. Access the accompanying disk. Click on the folder *sine8_intr* to Open (load) the file *sine8_intr.out*. This loads the executable file *sine8_intr.out* into the C6711 processor.
2. Select Debug → Run. Connect the OUT (connector J6) on the DSK board to a speaker or to an oscilloscope and verify the generation of a 1-kHz tone. The IN/OUT connectors (J7/J6) on the DSK board use a 3.5-mm jack audio cable.

The folder sine8_intr contains the necessary files to implement Example 1.1, which introduces some features of the tools.

1.4.2 Support Files

Create a new folder within your PC hard drive and name it *sine8_intr*. It is recommended that you place this folder in *c:\ti\myprojects* (it is assumed that you have installed CCS in *c:\ti*). Some of the same support files that are used in many examples in this book are included on the accompanying disk in the folder *Support*. For now, don't worry too much about the content or functions of these files. Additional support files are included in the CCS CD with the DSK package. Copy the following support files from the folder *Support* (on the accompanying disk) into the folder *sine8_intr* that you created in your hard drive:

1. C6xdsk.cmd: sample linker command file.
2. C6xdsk.h: header file that defines addresses of external memory interface, the serial ports, etc. (TI support file included with CCS).
3. C6xinterrupts.h: contains init functions for interrupt (TI support file included with the DSK).
4. C6xdskinit.h: header file with the function prototypes.
5. C6xdskinit.c: contains several functions used for the example codec_poll included with CCS. It includes functions to initialize the DSK, the codec, the serial ports, and for input/output.
6. Vectors_11.asm: version of vectors.asm included with CCS, but modified to handle interrupts. Twelve interrupts, INT4 through INT15, are available, and INT11 is selected within this vector file.

Also copy the C source file *sine8_intr.c* and the GEL file *amplitude.gel* from the disk (sine8_intr folder) into the folder *sine8_intr* on your hard drive.

Note: If you are using a C6211 DSK (which has been discontinued), change XINT0 to XINT1 within the function comm_intr in the file C6xdskinit.c. This is due to a silicon bug associated with the C6211.

1.4.3 Examples

Example 1.1: Sine Generation with Eight Points (sine8_intr)

This example generates a sinusoid using a table-lookup method. More important, it illustrates some features of CCS for editing, building a project, accessing the code generation tools, and running a program on the C6711 processor. The C source program *sine8_intr.c* shown in Figure 1.2 implements the sine generation.

```
//sine8_intr.c Sine generation using 8 points, f=Fs/(# of points)
//Comm routines and support files included in C6xdskinit.c

short loop = 0;
short sin_table[8] = {0,707,1000,707,0,-707,-1000,-707}; //sine values
short amplitude = 10;                       //gain factor

interrupt void c_int11()                    //interrupt service routine
{
 output_sample(sin_table[loop]*amplitude); //output each sine value
 if (loop < 7) ++loop;                      //increment index loop
 else loop = 0;                             //reinit index @ end of buffer
 return;                                    //return from interrupt
}

void main()
{
  comm_intr();                              //init DSK, codec, McBSP
  while(1);                                 //infinite loop
}
```

FIGURE 1.2. Sine generation program using eight points (sine8_intr.c).

Program Consideration

Although the focus is to illustrate some of the tools, it is useful to understand the program *sine8_intr.c*. A table or buffer sin_table is created and filled with eight points representing sin(t), where $t = 0, 45, 90, 135, 180, 225, 270$, and 315 degrees (scaled by 1000). Within the function *main*, another function *comm_intr* is called that is located in the communication support file *c6xdskinit.c*. It initializes the DSK, the AD535 codec onboard the DSK, and the two multichannel buffered serial ports (McBSPs) on the C6711 processor.

The statement while (1) within the function *main* creates an infinite loop to wait for an interrupt to occur. On interrupt, execution proceeds to the interrupt service routine (ISR) *c_int11*. This ISR address is specified in the file vectors_11.asm with a branch instruction to this address, using interrupt INT11. Interrupts are discussed in more detail in Chapter 3.

Within the ISR, the function output_sample, located in the communication support file *C6xdskinit.c*, is called to output the first data value in the buffer or table *sin_table[0]* = 0. The loop index is incremented until the end of the table is reached, after which case it is reinitialized to zero. Execution returns from ISR to the while(1) infinite loop to wait for the next interrupt to occur.

An interrupt occurs every sample period $T = 1/F_s = 1/8000 = 0.125$ ms. Every sample period 0.125 ms, an interrupt occurs, ISR is accessed, and a subsequent data value in sin_table (scaled by *amplitude* = 10) is sent for output. Within one period, eight data values (0.125 ms apart) are output to generate a sinusoidal signal.

The period of the output signal is $T = 8(0.125 \, \text{ms}) = 1 \, \text{ms}$, corresponding to a frequency of $f = 1/T = 1 \, \text{kHz}$.

Create Project

In this section we illustrate how to create a project, adding the necessary files for building the project **sine8_intr**. Access CCS (from the desktop).

1. To create the project file *sine8_intr.pjt*. Select Project → New. Type *sine8_intr* for project name as shown in Figure 1.3*a*. This project file is saved in *sine8_intr* (the folder you created in c:\ti\myprojects). The .pjt file stores project information on build options, source filenames, and dependencies.

2. To add files to project. Select Project → Add Files to Project. Look in *sine8_intr*, Files of type C Source Files. Open the two C source files *C6xdskinit.c* and *sine8_intr.c*. Open (to add to project) one file at a time; or place the cursor to one of these files, then to the other while holding the Shift key, and press Open. Click on the "+" symbol on the left of the Project Files window within CCS to expand and verify that the two C source files have been added to the project.

3. Select Project → Add Files to Project. Look in *sine8_intr*. Use the pull-down menu for Files of type: and select ASM Source Files. Double-click on the assembly source file *vectors_11.asm* to open/add it to the project.

4. Repeat step 3 but select Files of type: Linker Command File, and add the linker command file *C6xdsk.cmd* to the project.

5. Repeat step 3, but select Files of type: Object and Library Files. Look in c:\ti\c6000\cgtools\lib and select the run-time support library file *rts6701.lib* (which supports the C67x/C62x architecture) to add to the project. This assumes that you used the default destination of c:\ti when you installed CCS.

6. Verify that the linker command (.cmd) file, the project (.pjt) file, the library (.lib) file, the two C source (.c) files, and the assembly (.asm) file have been added to the project. The GEL file *dsk6211_6711.gel* is added automatically when you create the project. It initializes the DSK.

7. Note that there are no "include" files yet. Select Project → Scan All Dependencies. This adds/includes the header files: *C6xdsk.h*, *C6xdskinit.h*, *C6xinterrupts.h*, and *C6x.h*. The first three header files were copied (transferred) from the accompanying disk, and *C6x.h* is included with CCS.

The Files window in CCS should look as in Figure 1.3*b*. Any of the files (except the library file) from CCS's Files window can be displayed by clicking on it. You should not add header or include files to the project. They are added to the project automatically when you select: Scan All Dependencies.

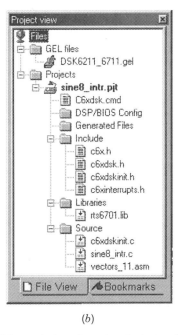

(a)

(b)

FIGURE 1.3. CCS Project View window for `sine8_intr`: (a) creating project; (b) project files.

It is also possible to add files to a project simply by "dragging" the file (from a different window) and dropping it into the CCS Project window.

Code Generation and Options
Various options are associated with the code generation tools: C compiler and linker to build a project.

Compiler Option. Select Project → Build Options. Figure 1.4*a* shows CCS window Build Options for the compiler. Select the following for the compiler option: (a) Basic (for Category), (b) Default (for Target Version), (c) Full Symbolic Debug (for Generate Debug Info), (d) Speed most critical (for Opt Speed vs. size), (e) None (for Opt Level and Program Level Opt). The resulting compiler option is

```
-gks
```

The $-k$ option is to keep the assembly source file *sine8_intr.asm*. The $-g$ option is to enable symbolic debugging information, useful during the debugging process, and used in conjunction with the option $-s$ to interlist the C source file with the assembly source file *sine8_intr.asm* generated. The $-g$ option disables many code optimizations to facilitate the debugging process.

Selecting "Default" for Target Version invokes a fixed-point implementation. (If you have a C6211 DSK, you must use this option.) The C6711-based DSK can use either fixed- or floating-point processing. Most examples implemented in this book can run using fixed-point processing. You will need to select C671x to invoke a floating-point implementation for the examples in Chapter 6 and 7.

If No Debug is selected (for Generate Debug Info), and $-o3:File$ is selected (for Opt Level), the Compiler option is automatically changed to

```
-ks -o3
```

The $-o3$ option invokes the highest level of optimization for performance or execution speed. For now, speed is not critical (neither is debugging). Use the compiler option $-gks$ (you can type it directly in the compiler command window). Initially, one would not optimize for speed but to facilitate debugging. There are a number of compiler options described in Ref. 26.

Linker Option. Click on Linker (from CCS Build Options) and select Absolute Executable (for Output Module), sine8_intr.out (for Output Filename), and Run-time Autoinitialization (for Autoinit Model). The output filename defaults to the name of the .pjt filename. The linker option should be displayed as in Figure 1.4(*b*)

```
-g -c -o "sine8_intr.out" -x
```

The $-c$ option is used to initialize variables at run time, and the $-o$ option is to name the linked executable output file sine8_intr.out. Press OK.

Note that you can choose to store the executable file within a subfolder "Debug," especially during the debugging stage of a project.

Again, these various options can be typed directly within the appropriate command windows.

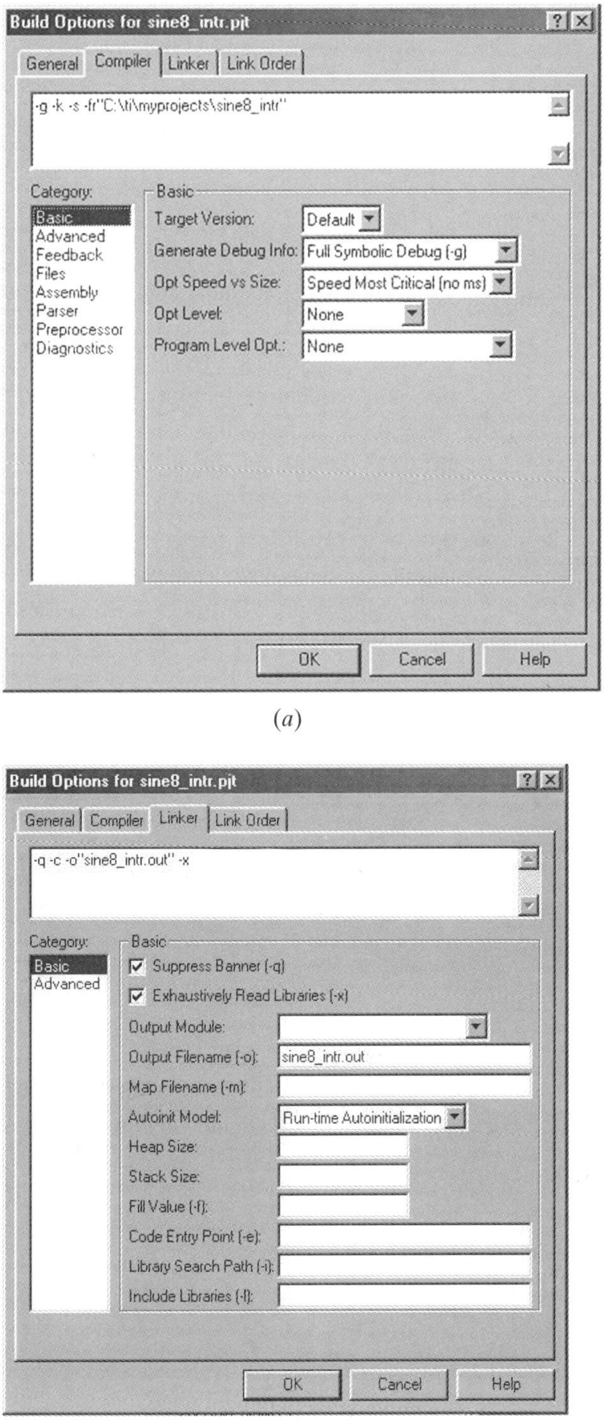

(*b*)

FIGURE 1.4. CCS Build options: (*a*) compiler; (*b*) linker.

Building and Running the Project

The project sine8_intr can now be built and run.

1. Build this project as **sine8_intr.** Select Project → Rebuild All. Or press the toolbar with the three down arrows. This compiles and assembles all the C files using cl6x and assembles the assembly file vectors_11.asm using asm6x. The resulting object files are then linked with the run-time library support file rts6701.lib using lnk6x. This creates an executable file sine8_intr.out that can be loaded into the C6711 processor and run. Note that the commands for compiling, assembling, and linking are performed with the Build option. A log file cc_build_Debug.log is created that shows the files that are compiled and assembled, along with the compiler options selected. It also lists the support functions that are used. Figure 1.5 shows several windows within CCS for the project sine8_intr.

2. Select File → Load Program in order to load sine_intr.out by clicking on it (CCS includes an option to load the program automatically after a build). It should be in the project sine8_intr folder. Select Debug → Run, or use the toolbar with the "running man." Connect a speaker to the OUT connector (J6) on the DSK. You should hear a tone.

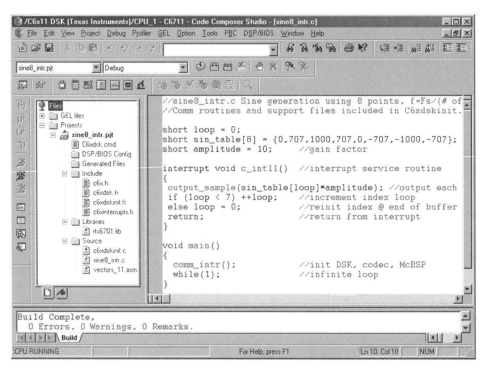

FIGURE 1.5. CCS windows for project sine8_intr.

The sampling rate F_s of the codec is fixed at 8 kHz. The frequency generated is $f = F_s/(\text{number of points}) = 8\,\text{kHz}/8 = 1\,\text{kHz}$. Connect the output of the DSK to an oscilloscope to verify a 1-kHz sinusoidal signal with an amplitude of approximately 0.85 V p-p (peak to peak).

Monitoring the Watch Window

Verify that the processor is still running. Note the indicator "DSP RUNNING" at the bottom left of CCS. The Watch window allows you to change the value of a parameter or to monitor a variable:

1. Select View → Quick Watch window, which should be displayed on the lower-section of CCS. Type `amplitude`, then click on "Add to Watch." The amplitude value of 10 set in the program in Figure 1.2 should appear in the Watch window.

2. Change `amplitude` from 10 to 30.

3. Verify that the volume of the generated tone has increased (note that the processor was still running). The amplitude of the sine wave has increased from approximately 0.85 V p-p to approximately 2.6 V p-p.

4. Change `amplitude` to 33 (as in step 2). Verify a higher-pitch tone, which implies that the frequency of the sine wave has changed just by changing its amplitude. This is not so. You have overflowed the capacity of the 16-bit codec AD535. Since the values in the table are scaled by 33, the range of these values is now between + and −33,000. The range of output values is limited from -2^{15} to $(2^{15} - 1)$, or from −32,768 to +32,767, due to the AD535 codec. Don't attempt to send more than 16 bits' worth of data to the codec. The onboard codec uses a 2's-complement format.

Correcting Program Errors

1. Delete the semicolon in the statement

   ```
   short amplitude = 10;
   ```

   If the C source file `sine8_intr` is not displayed, double-click on it (from the Files window).

2. Select Debug → Build to perform an incremental build or use the toolbar with the two (not three) arrows. The incremental build is chosen so that only the C source file `sine8_intr.c` is compiled. With the Rebuild option (toolbar with three arrows), files compiled and/or assembled previously would again go through this unnecessary process.

3. An error message, highlighted in red, stating that a ";" is expected, should appear in the Build window of CCS (lower left). You may need to scroll-up the Build window for a better display of this error message. Double-click on the highlighted error message line. This should bring the cursor to the section of code where the error occurs. Make the appropriate correction, Build again, Load, and Run the program to verify your previous results.

Applying the Slider Gel File

The General Extension Language (GEL) is an interpretive language similar to (a subset of) C. It allows you to change a variable such as amplitude, sliding through different values while the processor is running. All variables must first be defined in your program.

1. Select File → Load GEL and open the file `amplitude.gel`, that you copied (from the accompanying disk) into the folder `sine8_intr`. Double-click on the file *amplitude.gel* to view it within CCS. It should be displayed in the Files window. This file is shown in Figure 1.6. By creating the slider function *amplitude* shown in Figure 1.6, you can start with an initial value of 10 (first value) for the variable amplitude that is set in the C program, up to a value of 35 (second value), incremented by 5 (third value).

2. Select GEL → Sine Amplitude → Amplitude. This should bring out the Slider window shown in Figure 1.7, with the minimum value of 10 set for amplitude.

3. Press the up-arrow key to increase the amplitude value from 10 to 15, as displayed in the Slider window. Verify that the volume of the sine wave generated has increased. Press the up-arrow key again to continue increasing the slider, incrementing by 5 up to 30. The amplitude of the sine wave should be about 2.6 V p-p with an amplitude value set at 30. Now use the mouse to click on the Slider window and slowly increase the slider position to 31, then 32, and verify that the frequency generated is still 1 kHz. Increase the slider to 33 and verify that you are no longer generating a 1-kHz sine wave (rather a signal with two tones: 1 and 3 kHz). The table values, scaled by amplitude, are now between + and −33,000 (beyond the acceptable range by the codec).

Two sliders can readily be used, one to change the amplitude and the other to change the frequency. A different frequency can be generated by changing the loop index within the C program (e.g., stepping through every two points in the table; see Example 2.4). When you exit CCS after you build a project, all changes made to the project can be saved. You can later return to the project with the status as you left it before.

```
/*Amplitude.gel Create slider and vary amplitude of sinewave*/

menuitem "Sine Amplitude"

slider Amplitude(10,35,5,1,amplitudeparameter)   /*start at 10,up to 35*/
     {
         amplitude = amplitudeparameter;           /*vary amplit of sine*/
     }
```

FIGURE 1.6. GEL file to "slide" through different amplitude values in the sine generation program (`amplitude.gel`).

FIGURE 1.7. CCS slider window for varying the amplitude of a sine wave.

Example 1.2: Generation of Sinusoid and Plotting with CCS (*sine8_buf*)

This example generates a sinusoid with eight points, as in Example 1.1. More important, it illustrates CCS capabilities for plotting in both time and frequency domains. The program *sine8_buf.c* (Figure 1.8), implements this project. This program creates a buffer to store the output data in memory.

Create this project as *sine8_buf.pjt*, add the necessary files to the project as in Example 1.1 (use *sine8_buf.c* in lieu of *sine8_intr.c*). Note that the necessary header support files are added to the project by selecting Project → Scanning All Dependencies. All of the support files for this project are in the folder sine8_buf (on disk).

Build this project as **sine8_buf**. Load and run the executable file *sine8_buf.out* and verify that a 1-kHz sinusoid is generated with the output connected to a speaker or a scope (as in Example 1.1).

Plotting with CCS
The output buffer is being updated continuously every 256 points (you can readily change the buffer size). Use CCS to plot the current output data stored in the buffer *out_buffer*.

1. Select View → Graph → Time/Frequency.
2. Change the Graph Property Dialog so that the options in Figure 1.9*a* are selected for a time-domain plot (use the pull-down menu when appropriate). The starting address of the output buffer is out_buffer. The other options can be left as default. Figure 1.10 shows a time-domain plot of the sinusoidal signal.

```
//sine8_buf Sine generation. Output buffer plotted within CCS
//Comm routines and support files included in C6xdskinit.c

short loop = 0;
short sine_table[8] = {0,707,1000,707,0,-707,-1000,-707};  //sine values
short out_buffer[256];              //output buffer
const short BUFFERLENGTH = 256;     //size of output buffer
short i = 0;                        //for buffer count

interrupt void c_int11()                 //interrupt service routine
{
  output_sample(sine_table[loop]);  //output each sine value
  out_buffer[i] = sine_table[loop]; //output to buffer
  i++;                              //increment buffer count
  if (i == BUFFERLENGTH) i = 0;    //if bottom reinit buffer count
  if (loop < 7) ++loop;            //increment index loop
  else loop = 0;                   //if end of buffer,reinit index
  return;
}

void main()
{
  comm_intr();                      //init DSK, codec, McBSP
  while(1);                         //infinite loop
}
```

FIGURE 1.8. Sine generation with output stored in memory also (sine8_buf.c).

Graph Property Dialog	
Display Type	Single Time
Graph Title	Graphical Display
Start Address	out_buffer
Acquisition Buffer Size	256
Index Increment	1
Display Data Size	64
DSP Data Type	16-bit signed integer
Q-value	0
Sampling Rate (Hz)	8000

OK   Cancel   Help

(a)

Graph Property Dialog	
Display Type	FFT Magnitude
Graph Title	Graphical Display
Signal Type	Real
Start Address	out_buffer
Acquisition Buffer Size	256
Index Increment	1
FFT Framesize	256
FFT Order	8
FFT Windowing Funct	Rectangle
Display Peak and Ho	Off
DSP Data Type	16-bit signed integer
Q-value	0
Sampling Rate (Hz)	8000

OK   Cancel   Help

(b)

FIGURE 1.9. CCS Graph Property Dialog for sine8_buf: (a) for time-domain plot; (b) for frequency-domain plot.

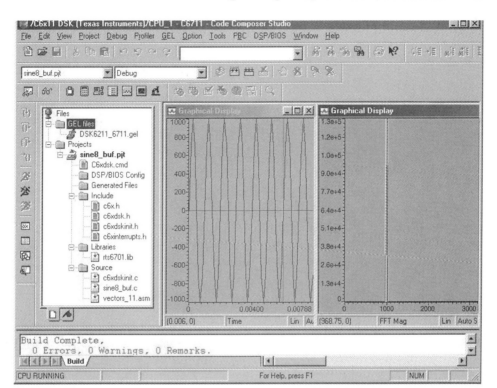

FIGURE 1.10. CCS windows with both time- and frequency-domain plots of a 1-kHz sine wave.

3. Figure 1.9*b* shows CCS's Graph Property Display for a frequency-domain plot. Choose an FFT order so that 2^{order} is the frame size. Press OK and verify that the FFT magnitude plot is as shown in Figure 1.10. The spike at 1000 Hz represents the frequency of the sinusoid generated.

Note: To change the screen size, right-click on the Build window and deselect Allow Docking. You can then obtain many different windows within CCS.

Example 1.3: Dot Product of Two Arrays (dotp4)

Operations such as addition/subtraction and multiplication are the key operations in a digital signal processor. A very important operation is the multiply/accumulate, which is useful in a number of applications requiring digital filtering, correlation, and spectrum analysis. Since the multiplication operation is executed so commonly and is so essential for most digital signal processing algorithms, it is important that it executes in a single cycle. With the C6x we can actually perform two multiply/accumulate operations within a single cycle.

This example illustrates additional features of CCS, such as single-stepping and profiling for benchmark. The focus here is to become still more familiar with the

```
//Dotp4.c Multiplies two arrays, each array with 4 numbers

int dotp(short *a, short *b, int ncount);       //function prototype
#include <stdio.h>                               //for printf
#include "dotp4.h"                               //data file of numbers
#define count 4                                  //# of data in each array
short x[count] = {x_array};                      //declara 1st array
short y[count] = {y_array};                      //declara 2nd array

main()
{
  int result = 0;                               //result sum of products

  result = dotp(x,y,count);                     //call dotp function
  printf("result = %d (decimal) \n", result);   //print result
}

int dotp(short *a, short *b, int ncount)        //dot product function
{
  int sum = 0;                                  //init sum
  int i;

  for (i = 0; i < ncount; i++)
    sum += a[i] * b[i];                         //sum of products
  return(sum);                                  //return sum as result
}
```

FIGURE 1.11. Sum-of-products program using C code (dotp4.c).

```
//dotp4.h Header file with two arrays of numbers

#define x_array 1,2,3,4

#define y_array 0,2,4,6
```

FIGURE 1.12. Header file with two arrays each with four numbers (dotp4.h).

tools. We invoke the C compiler optimization to see how performance or execution speed can be drastically increased.

The C source file dotp4.c (Figure 1.11) takes the sum of products of two arrays, each array with four numbers, contained in the header file dotp4.h (Figure 1.12). The first array contains the four numbers 1, 2, 3, and 4, and the second array contains the four numbers 0, 2, 4, and 6. The sum of products is $(1 \times 0) + (2 \times 2) + (3 \times 4) + (4 \times 6) = 40$.

The program can readily be modified to handle a larger set of data. No real-time implementation is used in this example, and no real-time I/O support files are

```
*Vectors.asm Vector file for non-interrupt driven program
     .title          "vectors.asm"
     .ref            _c_int00          ;reference entry address
     .sect           "vectors"         ;in vector section
rst: mvkl      .s2   _c_int00,b0       ;lower 16 bits -> b0
     mvkh      .s2   _c_int00,b0       ;higher 16 bits -> b0
     b         .s2   b0                ;branch to entry address
     nop                               ;5 NOPs for rest of fetch packet
     nop
     nop
     nop
     nop
```

FIGURE 1.13. Vector file for non-interrupt-driven program (`vectors.asm`).

needed. The support functions for interrupts are not needed here. The vector file used in this example is less extensive, as shown in Figure 1.13.

Create and build this project as **dotp4** and add the following files to the project as in Example 1.1:

1. dotp4.c: C source file
2. vectors.asm: vector file defining entry address *c_int00*
3. C6xdsk.cmd: linker command file
4. rts6701.lib: library file

Do not add any "include" files using "Add Files to Project" since they are added by selecting Project → Scan All Dependencies. The header file stdio.h is needed due to the printf statement in the program dotp4.c to print the result.

Implementing a Variable Watch

1. Select Project → Options:
   Compiler: -gs
   Linker: -c -o dotp4.out
2. Rebuild All by selecting the toolbar with the three arrows (or select Debug → Build).
3. Select View → Quick Watch. Type *sum* to watch the variable *sum*, and click on "Add to Watch." A message "identifier not found" associated with *sum* is displayed (as Value) because this local variable "does not exist" yet since we are still in the function *main*.
4. Set a breakpoint at the line of code

   ```
   sum += a[i] * b[i];
   ```

by placing the mouse cursor (clicking) on that line, then right-click and select Toggle breakpoint. A circle on the left of that line of code should appear.

5. Select Debug → Run (or use the "running man" toolbar). The program executes up to the line of code with the set breakpoint. A yellow arrow will also point to that line of code.

6. Single-step using F8 (or use the toolbar). Repeat or continue to single-step and observe/watch the variable *sum* change in value to 0, 4, 16, 40. Select Debug → Run, and verify that the resulting value of sum is printed as

```
sum = 40 (decimal)
```

7. Note the `printf` statement in the C program `dotp4.c` for printing the result. Such statement should be avoided, since it can take 3000 cycles to execute.

Animating

1. Select Debug → Rcset CPU → File → Reload Program to reload the executable file `dotp4.out`.

2. Again set the breakpoint as in the same line of code as before. Select Debug → Animate. Observe the variable *sum* change in values through the Watch window. The speed of animation can be controlled by selecting Option → Customize → Animate Speed.

Benchmarking without Optimization (Profiling)

In this section we illustrate how to benchmark a section of code: in this case, the *dotp* function. Verify that the same options for the compiler (-gs), and linker (-c -o dotp4.out) are still set. To profile code, you must use the compiler option -g for symbolic debugging information. Remove any breakpoint by clicking on the line of code with the breakpoint, right-click, and select Toggle breakpoint.

1. Select Debug → Reset CPU → File → Reload program, to reload the executable file.

2. Select Profiler → Start New Session, and enter `dotp4` as the Profile Session Name. Then press OK.

3. Click on the icon to "Create Profile Area" which is the fourth icon from the top left in Figure 1.14*b*. Figure 1.14*b* shows the added profile area for the function dotp within the C source file `dotp4.c`.

4. Run the program. Verify the results shown in Figure 1.14*b*. This indicates that it takes 138 cycles to execute the function dotp (with no optimization).

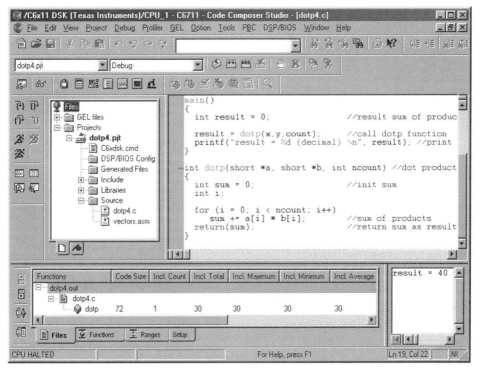

FIGURE 1.14. CCS display of project `dotp4` for profiling: (*a*) profile area of code lines 18–26; (*b*) profiling function dotp with no optimization; (*c*) profiling function dotp with optimization.

Benchmarking with Optimization (Profiling)

In this section we illustrate how to optimize using one of the optimization options −o3. The program's execution speed can be increased by the optimizing C compiler. Change the compiler option (select Project → Build Options) to

```
-g  -o3
```

and use the same linker options as before (you can type this option directly). The option −o3 invokes the highest level of compiler optimization. Various compiler options are described in Ref. 26. Rebuild All (toolbar with three arrows) and load the executable file (select File → Load Program) dotp4.out. Note that after the executable file is loaded, the entry address for execution is c_int00, as can be verified by the disassembled file.

Select Debug → Run. Verify that it takes now 30 cycles (from 138) to execute the dotp function, as shown in Figure 1.14c. This is a considerable improvement using the C compiler optimizer. We further optimize the dot product example using an intrinsic function in Chapter 3 and code optimization techniques in Chapter 8.

1.5 SUPPORT PROGRAMS/FILES CONSIDERATIONS

The following support files are used for practically all the examples in this book: (1) *C6xdskinit.c*, (2) *Vectors_11.asm*, and (3) *C6xdsk.cmd*. For now, the emphasis associated with these files should be on using them.

1.5.1 Initialization/Communication File (**C6xdskinit.c**)

The function *comm_intr* within main in the C source program is located in the communication file *c6xdskinit.c*, a partial listing of which is shown in Figure 1.15. The DSK is initialized, then the transmit interrupt INT11 is configured and enabled.

Two functions for input and output are also included in this communication support file. The function input_sample returns the input data value from *mcbsp0_read*, and the function output_sample calls *mcbsp0_write* for output.

Interrupt-Driven Program

With an interrupt-driven program, an interrupt is selected (we selected INT11). The nonmaskable interrupt bit must be enabled as well as the Global Interrupt Enable (GIE) bit. The appropriate support functions for interrupts are within the support file *C6xdskinterrupts.h* and are called from the function *comm_intr* within the file *C6xdskinit.c*.

```
//C6xdskinit.c Partial listing. Init DSK,AD535,McBSP

#include <c6x.h>
#include "c6xdsk.h"
#include "c6xdskinit.h"
#include "c6xinterrupts.h"

void mcbsp0_write(int out_data)        //function for writing
{
 int temp;

 if (polling)                          //bypass if interrupt-driven
 {
  temp = *(unsigned volatile int *)McBSP0_SPCR & 0x20000;
  while ( temp == 0)
     temp = *(unsigned volatile int *)McBSP0_SPCR & 0x20000;
 }
 *(unsigned volatile int *)McBSP0_DXR = out_data;
}

int mcbsp0_read()                      //function for reading
{
 int temp;

 if (polling)                          //bypass if interrupt-driven
 {
  temp = *(unsigned volatile int *)McBSP0_SPCR & 0x2;
  while ( temp == 0)
     temp = *(unsigned volatile int *)McBSP0_SPCR & 0x2;
 }
 temp = *(unsigned volatile int *)McBSP0_DRR;
 return temp;
}

void comm_poll()                       //communication with polling
{
   polling = 1;                        //setup for polling
   c6x_dsk_init();                     //call init DSK function
}

void comm_intr()                       //communication with interrupt
{
   polling = 0;                        //if interrupt-driven
   c6x_dsk_init();                     //call init DSK function
   config_Interrupt_Selector(11,XINT0); //using transmit interrupt INT11
   enableSpecificINT(11);             //for specific interrupt
   enableNMI();                        //enable NMI
   enableGlobalINT();                  //enable GIE global interrupt
   mcbsp0_write(0);                    //write to SP0
}

void output_sample(int out_data)       //added function for output
{
  mcbsp0_write(out_data & 0xfffe);     //mask out LSB
}

int input_sample()                     //added function for input
{
   return mcbsp0_read();               //read from McBSP0
}
```

FIGURE 1.15. Partial listing of communication support program (C6xdskinit.c).

Polling-Based Program

A polling-based program (non-interrupt driven) continuously polls or tests whether or not data are ready to be received or transmitted. This scheme is less efficient than the interrupt scheme. Within the input read function *mcbsp0_read*, the content of the serial port control register (SPCR) is ANDed with 0x2 to test bit 1 (second LSB) of the register, as shown in Figure B.8 (Appendix B). Within the output write function *mcbsp0_write*, SPCR is ANDed with 0x20000 to test bit 17. An input data value is accessed through the data receive register of the multichannel buffered serial port (McBSP). An output data value is sent through the data transmit register of McBSP.

We use the polling scheme later in several examples to control the input and output data rate. Most examples are interrupt driven. Interrupts are discussed in Chapter 3. For now, INT11 is generated via the serial port (McBSP).

1.5.2 Vector File (`vectors_11.asm`)

To select interrupt INT11, a branch instruction to the interrupt service routine (ISR) c_int11 located in the C program (sine8_intr.c or sine8_buf.c) is placed at the address INT11 in vectors_11.asm. A listing of the file vectors_11.asm is shown in Figure 1.16. Note the underscore preceding the name of the routine or function being called. The ISR is also referenced in vectors_11.asm using *.ref _c_int11*.

For a non-interrupt-driven vector program, modify vectors_11.asm:

1. Delete the reference to the interrupt service routine (ISR) *.ref _c_int11*.
2. For interrupt INT11, replace the branch instruction to the ISR by NOP.

1.5.3 Linker File (`C6xdsk.cmd`)

The linker command file *C6xdsk.cmd* is listed in Figure 1.17. It shows that sections such as .text and .stack reside in IRAM, which is mapped to the internal memory of the C6711 digital signal processor. It can be used as a generic sample linker command file even though some portion of it is not necessary. In Chapter 4 we show an example of the use of external RAM using SDRAM which starts at the address 0x80000000.

1.6 COMPILER/ASSEMBLER/LINKER SHELL

In previous examples the code generation tools for compiling, assembling, and linking were invoked within CCS while building a project. The tools may also be invoked directly outside CCS, using a DOS shell.

```
*Vectors_11.asm Vector file for interrupt-driven program
            .ref        _c_int11        ;ISR used in C program
            .ref        _c_int00        ;entry address
            .sect       "vectors"       ;section for vectors
RESET_RST:  mvkl   .S2  _c_int00,B0     ;lower 16 bits —> B0
            Mvkh   .S2  _c_int00,B0     ;upper 16 bits —> B0
            B      .S2  B0              ;branch to entry address
            NOP                         ;NOPs for remainder of FP
            NOP                         ;to fill 0x20 Bytes
            NOP
            NOP
            NOP
NMI_RST:    .loop 8
            NOP                         ;fill with 8 NOPs
            .endloop
RESV1:      .loop 8
            NOP
            .endloop
RESV2:      .loop 8
            NOP
            .endloop
INT4:       .loop 8
            NOP
            .endloop
INT5:       .loop 8
            NOP
            .endloop
INT6:       .loop 8
            NOP
            .endloop
INT7:       .loop 8
            NOP
            .endloop
INT8:       .loop 8
            NOP
            .endloop
INT9:       .loop 8
            NOP
            .endloop
INT10:      .loop 8
            NOP
            .endloop
INT11:      b        _c_int11           ;branch to ISR
            .loop 7
            NOP
            .endloop
INT12:      .loop 8
            NOP
            .endloop
INT13:      .loop 8
            NOP
            .endloop
INT14:      .loop 8
            NOP
            .endloop
INT15:      .loop 8
            NOP
            .endloop
```

FIGURE 1.16. Interrupt-driven vector program (vectors_11.asm).

```
/*C6xdsk.cmd Generic Linker command file*/

MEMORY
{
  VECS:        org =            0h, len =        0x220 /*vector section*/
  IRAM:        org = 0x00000220, len = 0x0000FDC0 /*internal memory*/
  SDRAM:       org = 0x80000000, len = 0x01000000 /*external memory*/
  FLASH:       org = 0x90000000, len = 0x00020000 /*flash memory*/
}

SECTIONS
{
  vectors   :> VECS
  .text     :> IRAM
  .bss      :> IRAM
  .cinit    :> IRAM
  .stack    :> IRAM
  .sysmem   :> SDRAM
  .const    :> IRAM
  .switch   :> IRAM
  .far      :> SDRAM
  .cio      :> SDRAM
}
```

FIGURE 1.17. Generic linker command file (C6xdsk.cmd).

1.6.1 Compiler

The compiler shell can be invoked using

```
Cl6x [options] [files]
```

to compile and assemble files that can be C files with extension .c, assembly files with extension .asm, and linear assembly (introduced in Chapter 3) with extension .sa. A linear assembly program file is a "cross" between C and assembly that can provide a compromise between the more versatile C program and the most efficient assembly program. For example,

```
Cl6x -gks -o3 file1.c, file2, file3.asm, file4.sa
```

invokes the C compiler to compile *file1* and *file2* (default to extension .c) and generates the assembly files *file1.asm* and *file2.asm*. This also invokes the assembler optimizer to optimize *file4.sa* and create *file4.asm*. Then the assembler (invoked with the shell command cl6x) assembles the four assembly source files and creates the four object files *file1.obj*, . . . , *file4.obj*. The option -gs adds debugger-specific information for debugging purposes and interlists C statements into assembly files, respectively. The −k option is to keep the assembly source files generated.

Four levels of compiler optimizations are available, with −o3 to invoke the highest level of optimization. Level 0 allocates variables to registers. Level 1 performs all level 0 optimizations and eliminates local common expressions and removes unused assignments. Level 2 performs all level 1 optimizations plus loop optimizations and rolling (discussed later). Level 3 performs all level 2 optimizations and removes functions that are not called. There are also compiler optimizations to minimize code size (with possible degradation in execution speed).

Note that full optimization may change memory locations that can affect the functionality of a program. In such cases, these memory locations must be declared as volatile. The compiler does not optimize volatile variables. A volatile variable is allocated to an uninitialized section in lieu of a register. Volatiles can be used when memory access is to be exactly as specified in the C code.

Initially, the functionality of a program is of primary importance. One should *not* invoke any (or too-high-level) optimization option initially while debugging, since additional debugger-specific information is provided to enhance the debugging process. Such additional information suppresses the level of performance. It is also difficult to debug a program after optimization since the lines of code are usually no longer arranged in a serial fashion. Compiler options can also be set using the environment variable with C_OPTION.

1.6.2 Assembler

An assembly file *file3.asm* can also be assembled using

```
asm6x file3.asm
```

to create file3.obj. The .asm extension is optional. The resulting object files must then be linked with a run-time support library to create an executable common object file format (COFF) file with extension .out that can be loaded directly and run on the digital signal processor.

1.6.3 Linker

The linker can be invoked using

```
lnk6x -c prog1.obj -o prog1.out -l rts6701.lib
```

The −c option tells the linker to use special conventions defined by the C environment for automatic variable initialization at run time (another linker option, −cr, initializes the variables at load time). The −l option invokes the run-time support library file `rts6701.lib`. These options [`-c` (or `-cr`) and `-l`] must be used when linking. The object file `prog1.obj` is linked with the library file and creates the executable file `prog1.out`. Without the `-o` option, the executable file `a.out` (by default) is created.

The linker can also be invoked with the compiler shell command with the `-z` option:

```
Cl6x -gks -o3 prog1.c prog2.asm -z -o prog.out -m prog.map
-l rts6701.lib
```

to create the executable file *prog.out*. The `-m` option creates a map file that provides a list of all the addresses of sections, symbols, and labels that can be useful for debugging.

Linker options include *-heap size* to specify the heap size in bytes for dynamic memory allocation (default is 1 kB) and the option *-stack size* to specify the C system stack size in bytes. Other linker options can be found in Ref. 24.

The linker allocates your program in memory using a default location algorithm. It places the various sections into appropriate memory locations, where code and data reside. By using a linker command file, with extension `.cmd`, one can customize the allocation process, specifying MEMORY and SECTIONS directives within the command file. The linker directive MEMORY (uppercase) defines a memory model and designates the origin and length of various available memory spaces. The directive SECTIONS (uppercase) allocate the output sections into defined memory and designate the various code sections to available memory spaces.

The sample linker command file, shown in Figure 1.17, can be used for almost all of the examples in the book. We will use internal memory (IRAM) for code and data. In Chapter 4 we illustrate implementation of a digital filter using external memory SDRAM, which starts at 0x80000000, with a length (size) of 0x1000000 = 16 MB. Flash starts at memory location 0x90000000 and has a length of 0x20000 = 128 kB.

The linker also links automatically *boot.obj* when using C programs to initialize the run-time environment, setting the entry point to *c_int00*. The symbol *_c_int00* is defined automatically when the linker option `-c` (or `-cr`) is invoked. The function *_c_int00*, included in the run-time support library, is the entry point in *boot.obj*, which sets up the stack and calls *main*. The run-time library support program *boot.c* is used to autoinitialize variables. The linker option `-c` invokes the initialization process with `boot.c`. Note that it is defined in the vector files *vectors_11.asm* and *vectors.asm*.

REFERENCES

Note: References 21 to 33 are included with the DSK package.

1. R. Chassaing, *Digital Signal Processing Laboratory Experiments Using C and the TMS320C31 DSK*, Wiley, New York, 1999.

2. R. Chassaing, *Digital Signal Processing with C and the TMS320C30*, Wiley, New York, 1992.

3. R. Chassaing and D. W. Horning, *Digital Signal Processing with the TMS320C25*, Wiley, New York, 1990.

4. N. Kehtarnavaz and M. Keramat, *DSP System Design Using the TMS320C6000*, Prentice Hall, Upper Saddle River, NJ, 2001.

5. N. Kehtarnavaz and B. Simsek, *C6x-Based Digital Signal Processing*, Prentice Hall, Upper Saddle River, NJ, 2000.

6. N. Dahnoun, *DSP Implementation Using the TMS320C6x Processors*, Prentice Hall, Upper Saddle River, NJ, 2000.

7. J. H. McClellan, R. W. Schafer, and M. A. Yoder, *DSP First: A Multimedia Approach*, Prentice Hall, Upper Saddle River, NJ, 1998.

8. C. Marven and G. Ewers, *A Simple Approach to Digital Signal Processing*, Wiley, New York, 1996.

9. J. Chen and H. V. Sorensen, *A Digital Signal Processing Laboratory Using the TMS320C30*, Prentice Hall, Upper Saddle River, NJ, 1997.

10. S. A. Tretter, *Communication System Design Using DSP Algorithms*, Plenum Press, New York, 1995.

11. A. Bateman and W. Yates, *Digital Signal Processing Design*, Computer Science Press, New York, 1991.

12. Y. Dote, *Servo Motor and Motion Control Using Digital Signal Processors*, Prentice Hall, Upper Saddle River, NJ, 1990.

13. J. Eyre, The newest breed trade off speed, energy consumption, and cost to vie for an ever bigger piece of the action, *IEEE Spectrum*, June 2001.

14. J. M. Rabaey, ed., VLSI design and implementation fuels the signal-processing revolution, *IEEE Signal Processing*, Jan. 1998.

15. P. Lapsley, J. Bier, A. Shoham, and E. Lee, *DSP Processor Fundamentals: Architectures and Features*, Berkeley Design Technology, Berkeley, CA, 1996.

16. R. M. Piedra and A. Fritsh, Digital signal processing comes of age, *IEEE Spectrum*, May 1996.

17. R. Chassaing, The need for a laboratory component in DSP education: a personal glimpse, *Digital Signal Processing*, Jan. 1993.

18. R. Chassaing, W. Anakwa, and A. Richardson, Real-time digital signal processing in education, *Proceedings of the 1993 International Conference on Acoustics, Speech and Signal Processing (ICASSP)*, Apr. 1993.

19. S. H. Leibson, DSP development software, *EDN Magazine*, Nov. 8, 1990.

20. D. W. Horning, An undergraduate digital signal processing laboratory, *Proceedings of the 1987 ASEE Annual Conference*, June 1987.

21. *TMS320C6000 Programmer's Guide*, SPRU198D, Texas Instruments, Dallas, TX, 2000.

22. *TMS320C6211 Fixed-Point Digital Signal Processor–TMS320C6711 Floating-Point Digital Signal Processor*, SPRS073C, Texas Instruments, Dallas, TX, 2000.

23. *TMS320C6000 CPU and Instruction Set Reference Guide*, SPRU189F, Texas Instruments, Dallas, TX, 2000.

24. *TMS320C6000 Assembly Language Tools User's Guide*, Texas Instruments, Dallas, TX, SPRU186G, 2000.

25. *TMS320C6000 Peripherals Reference Guide*, SPRU190D, Texas Instruments, Dallas, TX, 2001.

26. *TMS320C6000 Optimizing Compiler User's Guide*, SPRU187G, Texas Instruments, Dallas, TX, 2000.

27. *TMS320C6000 Technical Brief*, SPRU197D, Texas Instruments, Dallas, TX, 1999.

28. *TMS320C64x Technical Overview*, SPRU395, Texas Instruments, Dallas, TX, 2000.

29. *TMS320C6x Peripheral Support Library Programmer's Reference*, SPRU273B, Texas Instruments, Dallas, TX, 1998.

30. *Code Composer Studio User's Guide*, SPRU328B, Texas Instruments, Dallas, TX, 2000.

31. *Code Composer Studio Getting Started Guide*, SPRU509, Texas Instruments, Dallas, TX, 2001.

32. *TMS320C6000 Code Composer Studio Tutorial*, SPRU301C, Texas Instruments, Dallas, TX, 2000.

33. *TLC320AD535C/I Data Manual Dual Channel Voice/Data Codec*, SLAS202A, Texas Instruments, Dallas, TX, 1999.

34. B. W. Kernigan and D. M. Ritchie, *The C Programming Language*, Prentice Hall, Upper Saddle River, NJ, 1988.

35. *Details on Signal Processing* (quarterly publication), Texas Instruments, Dallas, TX.

36. G. R. Gircys, *Understanding and Using COFF*, O'Reilly & Associates, Newton, MA, 1988.

2

Input and Output with the DSK

- Input and output with the onboard AD535 codec (alternative input and output with the stereo codec PCM3003 are described in Appendix F)
- Programming examples using C code

2.1 INTRODUCTION

Typical applications using DSP techniques require at least the basic system shown in Figure 2.1, consisting of analog input and output. Along the input path is an antialiasing filter for eliminating frequencies above the *Nyquist frequency*, defined as one-half the sampling frequency F_s. Otherwise, aliasing occurs, in which case a signal with a frequency higher than one-half F_s is disguised as a signal with a lower frequency. The sampling theorem tells us that the sampling frequency must be at least twice the highest-frequency component f in a signal, so that

$$F_s > 2f$$

which is also

$$1/T_s > 2(1/T)$$

where T_s is the sampling period, or

$$T_s < T/2$$

The sampling period T_s must be less than one-half the period of the signal. For example, if we assume that the ear cannot detect frequencies above 20 kHz, we can

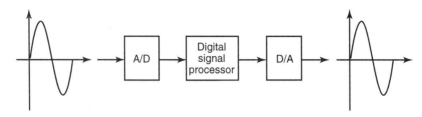

FIGURE 2.1. DSP system with input and output.

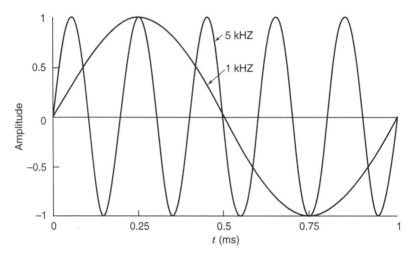

FIGURE 2.2. Aliased sinusoidal signal.

use a lowpass input filter with a bandwidth or cutoff frequency at 20 kHz to avoid aliasing. We can then sample a music signal at $F_s > 40$ kHz (typically, 44.1 kHz or 48 kHz) and remove frequency components higher than 20 kHz. Figure 2.2 illustrates an aliased signal. Let the sampling frequency $F_s = 4$ kHz, or a sampling period of $T_s = 0.25$ ms. It is impossible to determine whether it is the 5- or 1-kHz signal that is represented by the sequence (0, 1, 0, −1). A 5-kHz signal will appear as a 1-kHz signal; hence, the 1-kHz signal is an aliased signal. Similarly, a 9-kHz signal would also appear as a 1-kHz aliased signal.

2.2 TLC320AD535 (AD535) ONBOARD CODEC FOR INPUT AND OUTPUT

The DSK board includes the TLC320AD535 (AD535) codec for input and output. The ADC circuitry on the codec converts the input analog signal to a digital representation to be processed by the digital signal processor. The maximum level of the input signal to be converted is determined by the specific ADC circuitry on the codec, which is 3 V p-p with the onboard codec. After the captured signal is processed, the result needs to be sent to the outside world. Along the output path in Figure 2.1 is a DAC, which performs the reverse operation of the ADC. An output filter smooths out or reconstructs the output signal. ADC, DAC, and all required filtering functions are performed by the single-chip codec AD535 onboard the DSK.

The AD535 is a dual-channel voice/data codec based on sigma–delta technology [1–5]. It performs all the functions required for ADC and DAC, lowpass filtering, oversampling, and so on. The AD535 codec contains specifications for two channels and sampling rates of up to 11.025 kHz. However, the codec onboard the DSK has only one input and one output accessible readily by the user through two 3.5-mm audio cable connectors; and the sampling (conversion) rate is fixed at 8 kHz, not at 11.025 kHz [1].

Sigma–delta converters can achieve high resolution with high oversampling ratios but with lower sampling rates. They belong to a category where the sampling rate can be much higher than the Nyquist rate. The onboard AD535 codec over-samples by a factor of 64 times. A digital interpolation filter produces the over-sampling. The quantization noise power in such devices is independent of the sampling rate. A modulator is included to shape the noise so that it is spread beyond the range of interest. The noise spectrum is distributed between 0 and $F_s/2$, so that only a small amount of noise is within the signal frequency band. A digital filter is also included to remove the out-of-band noise.

The ADC converts an input signal into discrete output digital words in 2's-complement format that correspond to the analog signal value. The DAC includes an interpolation filter and a digital modulator. A decimation filter reduces the digital data rate to the sampling rate. The DAC's output is first passed through an internal lowpass reconstruction filter to produce an output analog signal. Low noise performance for both ADC and DAC is achieved using oversampling techniques with noise shaping provided by sigma–delta modulators.

The sampling rate F_s is set by the frequency of the codec master clock MCLK of 4096 kHz, such that

$$F_s = \mathrm{MCLK}/512 = 8\,\mathrm{kHz}$$

A diagram of the AD535 codec interfaced to the C6711 DSK is shown in Figure 2.3 and is included with the CCS package.

Serial communication techniques are used. Primary and secondary communications allow conversion of data and control transfer across the same serial port. A primary transfer is for data conversion, and a secondary transfer is for control. The least significant bit of a D/A data register is used for secondary communication request.

2.3 PCM3003 STEREO CODEC FOR INPUT AND OUTPUT

An audio daughter card based on the PCM3003 stereo codec is described in Appendix F [6]. Figure 2.4*a* shows a photo of the $3 \times 3\frac{1}{2}$ inch audio daughter card, and

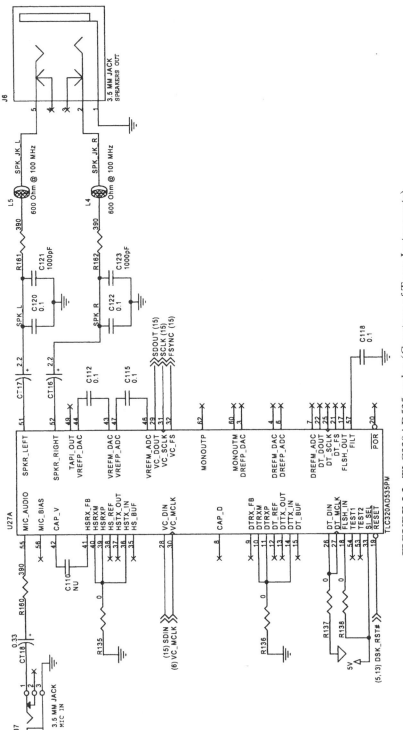

FIGURE 2.3. TLC320AD535 codec (Courtesy of Texas Instruments).

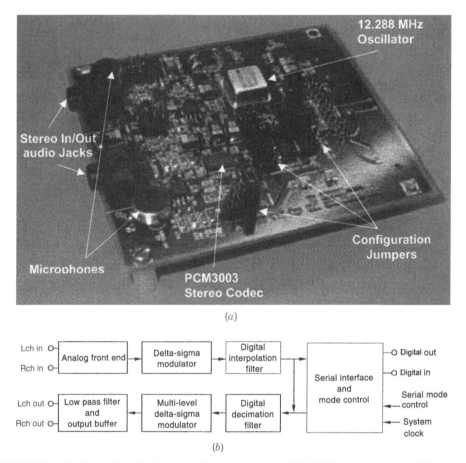

(a)

(b)

FIGURE 2.4. (*a*) Audio daughter card based on the PCM3003 stereo codec; (*b*) block diagram of PCM3003 codec (Courtesy of Texas Instruments).

Figure 2.4*b* shows a block diagram of the PCM3003 codec. A schematic for this daughter card is included in Appendix F. This daughter card plugs into the DSK through an 80-pin connector on the DSK board. The PCM3003 has two complete input and output channels and a variable programmable sampling rate with a maximum sampling rate of approximately 72 kHz (TI recommends a maximum of 48 kHz). Several programming examples using the PCM3003 are included in Appendix F to illustrate the use of a stereo codec with two input and output channels.

2.4 PROGRAMMING EXAMPLES USING C CODE

Several examples follow to illustrate input and output with the DSK. They are included to become more familiar with both the hardware and software tools and can provide some background to implement a specific application. For example, the project (example) `sine2sliders` illustrates the use of two sliders, an echo project demonstrates the effects of a variable-length buffer on an echo, an alternative echo project illustrates the use of two interrupts, and a square-wave generation project generates a square wave and illustrates how the AD535 translates a value to a corresponding output voltage. A list of all the examples included in this book appears on pages xv–xviii.

Example 2.1: Loop Program Using Interrupt (loop_intr)

This example illustrates input and output with the AD535 codec. Figure 2.5 shows the C source program *loop_intr.c*, which implements the loop program. It is interrupt-driven using INT11, as in Example 1.1.

This program example is very important since it can be used as a base program to build on. For example, to implement a digital filter, one would need to insert the appropriate algorithm between the "input" and "output" functions. The two functions input_sample and output_sample as well as the function comm_intr are included in the communication support file *C6xdskinit.c*. This is done so that the C source program is kept as small as possible. The file *C6xdskinit.c* can be used as a "black box program" since it is used in many examples throughout this book.

After the initialization and selection/enabling of an interrupt, execution waits within the infinite while loop until an interrupt occurs. Upon interrupt, execution proceeds to the interrupt service routine (ISR) c_int11, as specified in the vector file vectors_11.asm. An interrupt occurs every sample period $T_s = 1/F_s = 1/(8\,\text{kHz}) = 0.125\,\text{ms}$, at which time an input sample value is read from the codec's ADC, then sent as output to the codec's DAC.

```
//Loop_intr.c Loop program using interrupt, output is delayed input
//Comm routines and support files included in C6xdskinit.c

interrupt void c_int11()          //interrupt service routine
{
   int sample_data;

   sample_data = input_sample(); //input data
   output_sample(sample_data);    //output data
   return;
}

void main()
{
  comm_intr();                     //init DSK, codec, McBSP
  while(1);                        //infinite loop
}
```

FIGURE 2.5. Loop program using interrupt (loop_intr.c).

Execution returns from interrupt to the while(1) statement waiting for a subsequent interrupt. [Note that in lieu of waiting within the while(1) infinite loop, one could be processing code.] Upon interrupt, execution proceeds to ISR, "services" the necessary task dictated by ISR, then returns to the calling function waiting for the occurrence of a subsequent interrupt.

1. Within the function *output_sample*, the least signigficant bit of the output data value is masked for secondary communication or transfer. The DAC in the AD535 codec is effectively a 15-bit device since it uses the 15 MSBs of a 16-bit word as output data and the least significant bit (LSB) for control purposes. Within the function *output_sample*, the LSB of the 16-bit output data value is masked off, signaling the codec not to expect subsequent control data.

2. Within the function *comm_intr*, the following tasks are performed.

   (a) Initialize the DSK.

   (b) Configure/select INT11 and transmit interrupt XINT0.

   (c) Enable the specific interrupt.

   (d) Enable the global enable interrupt (GIE) bit.

   (e) Access the multichannel buffered serial port (McBSP) zero.

The interrupt functions called for the tasks above are included in the file *C6xinterrupts.h*, included with CCS.

Create and build this project as **loop_intr**. Use the same support files as in Example 1.1. All the source files used in this book and some support files are included on the accompanying disk. Other needed support files are included with CCS. Input a sinusoidal waveform to the IN connector J7 on the DSK, with an amplitude of approximately 1 to 2 V p-p and a frequency between approximately 1 and 3 kHz. Connect the output of the DSK, OUT of connector J6, and verify a tone of the same input frequency, with a small decrease in amplitude. Using an oscilloscope, the output is a delayed version of the input signal. Increase the amplitude of the input sinusoidal waveform beyond 3 V p-p and observe that the output signal becomes distorted.

Example 2.2: Loop Program Using Polling (loop_poll)

This example implements a loop program using polling to input and output a sample value every sample period T_s, whereas the program loop_intr.c in Example 2.1 is an interrupt-driven program. The C source program loop_poll.c (Figure 2.6) implements this loop program. The polling technique uses a continuous procedure of testing when the data are ready. Although it is simpler than the interrupt technique, it is less efficient.

1. Within the function input_sample, another function, *mcbsp0_read*, is called to read the input to the ADC from the data receive register (DRR) of

```
//loop_poll.c Loop program using polling, output is delayed input
//Comm routines and support files included in C6xdskinit.c

void main()
{
  int sample_data;

  comm_poll();                          //init DSK, codec, McBSP
  while(1)                              //infinite loop
  {
    sample_data = input_sample();  //input sample
    output_sample(sample_data);    //output sample
  }
}
```

FIGURE 2.6. Loop program using polling (loop_poll.c).

the multichannel buffered serial port (McBSP) 0, or simply SP0. The serial port control register (SPCR) is first ANDed with 0x2 to test if the receive ready register (RRDY) bit 1 of SPCR is enabled, as shown in Figure B.8.

2. Within the function output_sample, another function, *mcbsp0_write*, is called to write the output from the DAC to the data transmit register (DXR) of the McBSP 0 (SP0). SPCR is first ANDed with 0x20000 to test if the transmit ready register (XRDY) bit 17 of SPCR is enabled. Execution again waits within the infinite while(1) loop until the data are ready for transfer. At that time execution proceeds to input a sample data value and then output it.

The same support files are used as those in Example 2.1 or 1.1 except for the vector file vectors_11.asm. You can either replace vectors_11.asm (which uses INT11) with the file vectors.asm (on disk) or edit the file vectors_11.asm:

1. Delete .ref _c_int11, which is the assembler directive that references the interrupt service routine (ISR) _c_int11. The first underscore is the convention used with C functions.
2. Replace the instruction: b _c_int11, which is to branch to ISR, by a NOP (no operation).

Create and build this project as **loop_poll**. Use the same input as in Example 2.1, and verify the same results.

Example 2.3: Sine Generation Using Polling (sine4_poll)

This example generates a sinusoidal waveform using four points to further illustrate the use of polling. Figure 2.7 shows the C source program sine4_poll.c that implements the sine generation project with four points.

```
//Sine4_poll.c Sine generation using 4 points; f=Fs/(# points)=2 kHz

int loop = 0;
short sine_table[4] = {0,1000,0,-1000};          //sine values
short amplitude = 1;                             //for slider

void main()
{
  int sample_data;

  comm_poll();                                   //init DSK, codec, McBSP
  while(1)                                        //infinite loop
  {
   sample_data = (sine_table[loop]*amplitude);   //scaled value
   output_sample(sample_data);                   //output sine value
   if (loop < 3) ++loop;                          //increment index
   else loop = 0;                                 //reinit @ end of buffer
  }
}
```

FIGURE 2.7. Sine generation program using four points with polling (sine4_poll.c).

Use the same support file as with loop_poll in Example 2.2 (see also Example 1.1). At each sample period $T_s = 1/F_s$, the output consists of a data value from the buffer (table) sine_table. The data values 0, 1000, 0, −1000, 0, 1000, . . . are sent for output every 0.125 ms.

Build and run this project as **sine4_poll**. Verify that the output is a sine waveform with a dc offset of about 1 V due to the AD535 codec. The frequency generated is $f = F_s/(\text{number of points}) = 8\,\text{kHz}/4 = 2\,\text{kHz}$.

Load the GEL file *sine4_poll.gel* (Figure 2.8) and access the slider function amplitude as in Example 1.1. Change the slider from position 1 to positions 2, 3, . . . , 10 and verify the increase in amplitude (volume) of the waveform signal.

Change the slider function amplitude to start at 30 and up to 90 (in lieu of 10), still incrementing by 1. You can edit the GEL file, save it as sine4_poll.gel, reload, and access it through GEL. When the slider is at position 32, the output

```
/*Sine4_poll.gel Create slider and vary amplitude of sine wave*/

menuitem "Sine Amplitude"

slider Amplitude(1,10,1,1,amplitudeparameter) /*incr by 1,up to 10*/
{
     amplitude = amplitudeparameter;         /*vary amplitude of sine*/
}
```

FIGURE 2.8. GEL file to illustrate slider function (sine4_poll.gel).

```
//Sine2sliders.c Sine generation with different # of points

short loop = 0;
short sine_table[32]={0,195,383,556,707,831,924,981,1000,
                      981,924,831,707,556,383,195,
                      0,-195,-383,-556,-707,-831,-924,-981,-1000,
                   -981,-924,-831,-707,-556,-383,-195}; // sine data
short amplitude = 1;                          //for slider
short frequency = 2;                          //for slider

void main()
{
 comm_poll();                                 //init DSK, codec, McBSP
 while(1)                                     //infinite loop
   {
    output_sample(sine_table[loop]*amplitude); //output scaled value
    loop += frequency;                         //incr frequency index
    loop = loop % 32;                          //modulo 32 to reset
   }
}
```

FIGURE 2.9. Sine generation making use of two sliders for control of the amplitude and frequency generated (sine2sliders.c).

amplitude voltage is approximately 2.7 V p-p, with the sine values at + and −32,000. Increase the slider to 33, 34, . . . , 65, and observe that the amplitude *decreases* to about 0.1 V p-p with the slider at position 65. Does the amplitude of the waveform start to increase again with the slider at position 66, 67, . . . , 90?

Example 2.4: Sine Generation with Two Sliders for Amplitude and Frequency Control (sine2sliders)

The program *sine2sliders.c* (Figure 2.9) generates a sine wave using polling to control the output rate. Two sliders are used to vary both the amplitude and the frequency of the sinusoid generated. Using a lookup table with 32 points, the variable frequency is obtained by selecting different number of points per cycle. The amplitude slider scales the volume/amplitude of the waveform signal. The appropriate GEL file *sine2sliders.gel* is shown in Figure 2.10.

The 32 sine data values in the table or buffer correspond to sin(t), where $t = 0$, 11.25, 22.5, 33.75, 45, . . . , 348.75 degrees (scaled by 1000). The frequency slider takes on the values from 2 to 8, incremented by 2. The modulo operator is used to test when the end of the buffer that contains the sine data values is reached. When the loop index reaches 32, it is reinitialized to zero. For example, with the frequency slider at position 2, the loop or frequency index steps through every other value in the table. This corresponds to 16 data values within one cycle.

```
/*Sine2sliders.gel Two sliders to vary amplitude and frequency*/

menuitem "Sine Parameters"

slider Amplitude(1,8,1,1,amplitudeparameter) /*incr by 1,up to 8*/
    {
       amplitude = amplitudeparameter;        /*vary amplitude*/
    }

slider Frequency(2,8,2,2,frequencyparameter) /*incr by 2,up to 8*/
    {
       frequency = frequencyparameter;        /*vary frequency*/
    }
```

FIGURE 2.10. GEL file with two slider functions to control amplitude and frequency of the sine wave generated (`sine2sliders.gel`).

Build this project as **sine2sliders**. Use the same support files as in Example 2.3. Verify that the frequency generated is $f = F_s/16 = 500\,Hz$. Increase the slider position to 4, 6, 8, and verify that the signal frequencies generated are 1000, 1500, and 2000 Hz, respectively. Note that when the slider is at position 4, the loop or frequency index steps through the table selecting the eight values (per cycle): sin[0], sin[4], sin[8], ..., sin[28], that correspond to the data values 0, 707, 1000, 707, 0, −707, −1000, and −707. The resulting frequency generated is $f = F_s/8 = 1\,kHz$ (as in Example 1.1).

Example 2.5: Loop Program with Input Data Stored in Memory Buffer (loop_store)

The program *loop_store.c* (Figure 2.11) is an interrupt-based program. Each time an interrupt INT11 occurs, a sample is read from the codec's ADC and written to the codec's DAC. Furthermore, each sample is written to a 512-element circular buffer implemented using an array *buffer* and an index `i` that is incremented after each sample is stored. The index is reset to zero when it is incremented to 512. Consequently, the array always contains the 512 most recent sample values.

Build this project as **loop_store**. Input a sinusoidal signal with an amplitude of approximately $\frac{1}{2}$V p-p and a frequency of 1 kHz. Run and verify your output results.

Use CCS to plot the input data, in both the time and frequency domains (see also Example 1.2). Select View → Graph → Time/Frequency. Use a starting address "buffer" and chose 128 points (in lieu of 512 points) for the display data size to get a clearer plot, as shown in the Graph Property Dialog in Figure 2.12a (use other entries as default). Verify the 1-kHz time-domain sine-wave plot within CCS, as shown in Figure 2.12b.

Right-click on the graph window, or again, select View → Graph → Time/Frequency. Select FFT magnitude for display, as shown in the Graph Property Dialog

```
//Loop_store.c Data acquisition. Input data also stored in buffer

#define BUFFER_SIZE 512          //buffer size
short buffer[BUFFER_SIZE];       //buffer buffer
short i = 0;

interrupt void c_int11()         //interrupt service routine
{
 int sample_data;

 sample_data = input_sample(); //new input data
 output_sample(sample_data);   //output data

 buffer[i] = sample_data;       //store data in buffer
 i++;                           //increment buffer index
 if (i == BUFFER_SIZE) i = 0;   //reinit index if buffer full
 return;                        //return from ISR
}

void main()
{
 comm_intr();                   //init DSK, codec, McBSP
 while(1);                      //infinite loop
}
```

FIGURE 2.11. Loop program with input/output data in memory (loop_store.c).

in Figure 2.12c to obtain a frequency-domain plot of the input data. Note that the FFT order is $M = 9$, where $2^M = 512$. The spike at 1 kHz in Figure 2.12d represents the 1-kHz sine wave.

Example 2.6: Loop with Data in Buffer Printed to File (*loop_print*)

This example extends the preceding loop program so that the input/output data stored in a memory buffer are printed into a file. Figure 2.13 shows the C source program *loop_print.c* that implements this project example. It takes a long time (on the order of 4000 cycles) to execute the printf statement in the program. This can be reduced to about 30 cycles using real-time data transfer (RTDX), introduced in Appendix G.

After initialization of the DSK, the *puts* statement prints the word start as an indicator, then execution proceeds to the infinite while loop. Upon each interrupt, execution proceeds to ISR, and a newly acquired data value is stored into a buffer of size 64.

The buffer index i is incremented to store each new sampled data value. When

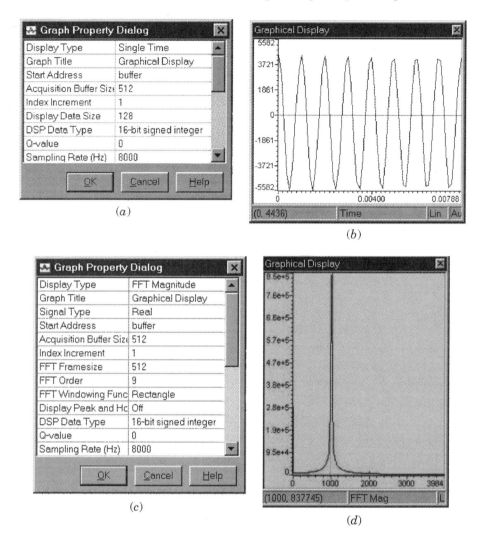

FIGURE 2.12. CCS graphs for `loop_store` program: (*a*) Graph Property Dialog displaying parameters for time-domain plot; (*b*) time-domain plot of stored output data representing 1-kHz sine wave; (*c*) Graph Property Dialog displaying parameters for FFT magnitude plot; (*d*) FFT magnitude of stored output data representing 1-kHz sine wave.

the end of the buffer is reached, indicating that the buffer is full, a file `loop.dat` is "opened" and the content of the buffer are written into that file. Then the indicator `done` is printed within the CCS command window. This process is repeated continuously so that a new set of 64 data points is acquired, and the `done` indicator is again displayed (after each set of data fills the buffer and written to `loop.dat`).

Build and run this project as ***loop_print***. Input a sine-wave signal of 1 V p-p

//Loop_print.c Data acquisition. Loop with data printed to a file

```
#include <stdio.h>
#define BUFFER_SIZE 64                      //buffer size
int i=0;
int j=0;
int buffer[BUFFER_SIZE];                    //buffer for data
FILE *fptr;                                  //file pointer

interrupt void c_int11()                     //interrupt service routine
{
 int sample_data;

 sample_data = input_sample();              //new input data
 buffer[i] = sample_data;                   //store data in buffer
 i++;                                        //increment buffer count
 if (i == BUFFER_SIZE - 1)                  //if buffer full
  {
   fptr = fopen("loop.dat","w");            //create output data file
   for (j=0; j<BUFFER_SIZE; j++)
     fprintf(fptr,"%d\n", buffer[j]);       //write buffer data to file
   fclose(fptr);                             //close file
   i = 0;                                    //initialize buffer count
   puts("done");                             //finished storing to file
  }
 output_sample(sample_data);                //output data
 return;                                     //return from ISR
}

void main()
{
 comm_intr();                                //init DSK, codec, McBSP
 puts("start\n");                            //print "start" indicator
 while(1);                                   //infinite loop
}
```

FIGURE 2.13. Loop program to store input/output data in memory and into a file
(loop_print.c).

with a 1-kHz frequency. Halt execution after the indicator done is displayed. The buffer of 64 input data representing the sine wave can be retrieved from the file *loop.dat*. Note that the third set of 64 points would be stored in the buffer and printed in the file *loop.dat* if execution is halted after the third done indicator. You can then use a plot program or MATLAB to plot loop.dat, and verify a 1-kHz sine wave. The output will not be displayed appropriately in real time, due to the slow execution of the print statements.

Example 2.7: Square-Wave Generation Using Lookup Table (squarewave)

This example generates a square wave using a lookup table and illustrates the data format of the AD535 codec. Figure 2.14 shows a listing of the program *square-wave.c* that implements this project example. A buffer of size 256 is created. Within *main*, the buffer table is loaded with data: the first half with $(2^{15} - 1) = 32{,}767$ and the second half with $-2^{15} = -32{,}768$. Upon each interrupt that occurs every sample period T_s, one data value from the buffer is sent for output. After each data value from the table is output, execution returns to the infinite while loop, waiting for the next interrupt to occur and output the subsequent value in the table. When the end of the buffer (table) is reached, the buffer index is reinitialized to the beginning of the buffer.

Build and run this project as **squarewave**. Verify a square-wave output signal of 2.8 V p-p with an offset of approximately 1.1 V.

Note that due to the 16-bit codec, the valid input data to the codec are between -2^{15} and $(2^{15} - 1)$ or between $-32{,}768$ and $32{,}767$. Change the values in the first half of the table using 0x8000 = 32,768 in lieu of 0x7FFF = 32,767. Rebuild/run and verify that a square-wave signal is no longer generated.

```
//Squarewave.c Generates a squarewave using a look-up table

#define table_size (int)0x100          //size of table = 256
int data_table[table_size];            //data table array
int i;

interrupt void c_int11()               //interrupt service routine
{
output_sample(data_table[i]);          //output value each Ts
if (i < table_size) ++i;               //if table size is reached
else i = 0;                            //reinitialize counter
return;                                //return from interrupt
}

main()
{
for(i=0; i<table_size/2; i++)          //set 1st half of buffer
     data_table[i] = 0x7FFF;           //with max value (2^15)-1
for(i=table_size/2; i<table_size; i++) //set 2nd half of buffer
     data_table[i] = -0x8000;          //with -(2^15)

i = 0;                                 //reinit counter
comm_intr();                           //init DSK, codec, McBSP
while (1);                             //infinite loop
}
```

FIGURE 2.14. Square-wave generation program (squarewave.c).

```
//Ramptable.c Generates a ramp using a look-up table

#define table_size (int)0x400    //size of table=1024
int data_table[table_size];     //data table array
int i;

interrupt void c_int11()        //interrupt service routine
{
 output_sample(data_table[i]);  //ramp value for each Ts
 if (i < table_size-1) i++;     //if table size is reached
 else i = 0;                    //reinitialize counter
 return;                        //return from interrupt
}

main()
{
  for(i=0; i < table_size; i++)
   {
    data_table[i] = 0x0;        //clear each buffer location
    data_table[i] = i * 0x20;   //set to 0,32,64,96, ... ,32736
   }
  i = 0;                        //reinit counter
  comm_intr();                  //init DSK, codec, McBSP
  while (1);                    //infinite loop
}
```

FIGURE 2.15. Ramp generation program using a table lookup (ramptable.c).

Example 2.8: Ramp Generation Using Lookup Table (ramptable)

Figure 2.15 shows a listing of the program ramptable.c, which generates a ramp using a lookup table. A buffer of size 1024 is created. Within main, the buffer table is loaded with 1024 values: 0, 0x20, 0x40, ..., or 0, 32, 64, ..., 32,736 in decimal.

Build and run this project as **ramptable**. Verify that a ramp is generated. The ramp's peak value is at the offset of approximately. 1.1 V and *decreases* for the input values 32, 64, ..., due to the 2's-complement format of the AD535 codec. As a result the ramp generated has a negative slope, with a peak-to-peak value of approximately 1.4 V.

Replace the value 0x20 with −0x20 and verify that a ramp is generated with a positive slope with a peak-to-peak value of 1.4 V. The ramp starts at the offset value of approximately 1.1 V and increases to approximately 2.5 V.

Example 2.9: Ramp Generation without a Lookup Table (ramp)

Example 2.8 is based on loading a table with a set of values, then outputting each value in the table every sample period, wrapping around when the end of the table

```
//Ramp.c Generates a ramp

int output;

interrupt void c_int11()   //interrupt service routine
{
 output_sample(output);    //output for each sample period
 output += 0x20;           //incr output value
 if (output == 0x8000)     //if peak is reached
 output = 0;               //reinitialize
 return;                   //return from interrupt
}

 void main()
{
 output = 0;               //init output to zero
 comm_intr();              //init DSK, codec, McBSP
 while(1);                 //infinite loop
}
```

FIGURE 2.16. Ramp generation program (`ramp.c`).

is reached. Figure 2.16 shows a listing of the program `ramp.c`, which generates a ramp using an alternative approach to Example 2.8. Starting with an initial output value of 0, the output value is incremented by `0x20` every sample period T_s. The values sent for output are then 0, 32, 64, 96, . . . , 32,736.

Build and run this project as **ramp**. Verify the same results as in Example 2.8, yielding a ramp with a negative slope.

To obtain a ramp with a positive slope, change output to

```
output  -= 0x20;
```

so that the output becomes 0, –32, –64, . . . , –32,736. Also change the if statement to reinitialize output, or

```
if (output  ==  -0x8000)
```

Verify that the output is a ramp with a positive slope.

Example 2.10: Echo (echo)

Figure 2.17 shows a listing of the program `echo.c`, which echoes an input signal. The length or size of the buffer determines the echo effect. A buffer size of 2000 barely generates a clear echo, while a size of 16,000 produces too much delay and the effect is more of a repeat. The output consists of a newly acquired sample added

```
//Echo.c Echo effect changed with size of buffer (delay)
short input, output;
short bufferlength = 3000;                //buffer size for delay
short buffer[3000];                       //create buffer
short i = 0;
short amplitude = 5;                      //to vary amplitude of echo

interrupt void c_int11()                  //ISR
{
 input = input_sample();                  //newest input sample data
 output=input + 0.1*amplitude*buffer[i];  //newest sample+oldest sample
 output_sample(output);                   //output sample

 buffer[i] = input;                       //store newest input sample
 i++;                                     //increment buffer count
 if (i >= bufferlength) i = 0;            //if end of buffer reinit
}

main()
{
 comm_intr();                             //init DSK, codec, McBSP
 while(1);                                //infinite loop
}
```

FIGURE 2.17. Echo generation (echo.c).

to the oldest sample already stored in the buffer. If the buffer size is too small, the time delay between the newest and oldest sample is too small to create an audible echo effect. The oldest sample is attenuated to enhance the echo effect.

After a new sample is acquired and stored at memory location x, the output becomes the sum of the new sample and the oldest sample stored at memory location x + 1, where x = 0, 1, 2, . . . , 2998. When the buffer index reaches the end of the buffer (buffer[2999]), where a newly acquired sample is stored, the oldest sample is at the beginning of the buffer.

Build and run this project as **echo**. A wave file, *Theforce.wav* (included on the accompanying disk), can be used as input. Play this file continuously with loop-around. The shareware utility Goldwave (described in Appendix E) allows you to play this file.

Change the size of the buffer from 1000 to 8000 and observe that a larger buffer size produces a greater delay between the newest and oldest samples. A GEL file (on the disk) can be used to increase or decrease the amplitude or effect of the echo.

A fading effect is obtained if the output (in lieu of the input) is stored in the buffer, using

```
buffer[i] = output;
```

Rebuild/run and verify this fading echo effect.

***Example 2.11: Echo Using Two Interrupts with Control for Different Effects
(echo_control)***

This example extends Example 2.10 to incorporate additional echo effects. It uses an alternative approach with two interrupts for reading/writing. Three sliders are used to vary the amplitude of the oldest sample, to change the buffer size for different amount of delay and to create a fading effect. The program *echo_control.c*, listed in Figure 2.18, implements this project. It uses the transmit interrupt INT11 to write as in earlier examples. In addition, it also uses the receive interrupt INT12 to read. The support files are the same as in previous examples, except:

1. The file *C6xdskinit.c* is modified to handle two interrupts. The following two lines of code are added in C6xdskinit.c:

```
config_Interrupt_Selector(12,RINT0); //receive INT12

enableSpecificInt(12);                //interrupt 12
```

2. The vector file *vectors_11.asm* is modified. Associated with INT12, add a branch statement to the interrupt service routine *c_int12*. This change is incorporated in the file vectors_11_12.asm.

Use the same .wav file, *Theforce.wav* (on the disk), for input as in Example 2.10. At each sample period, an interrupt occurs: first INT11 for writing, then INT12 for reading. The output is the sum of the newest input sample plus the oldest sample.

1. Build and run this project as **echo_control**.
2. Access the three sliders: amplitude, delay, and type. The GEL file *echo_control.gel* is shown in Figure 2.19. Set the amplitude slider to position 5, and set the delay slider to position 3. Since delay is not equal to delay_flag, the size of the buffer has changed. The new buffer size is buffer-length = $1000 \times 3 = 3000$. These two slider settings correspond to the same conditions as in Example 2.10. The delay slider can take on the values 1, 2, . . . , 8, allowing for buffer lengths of 1000, 2000, 3000, . . . , 8000. Increase the delay slider to position 4, then position 5, to produce a longer time delay between the newest and oldest samples and observe the echo effects.
3. The slider "type" in position 1 creates/adds a fading effect, since the output becomes the most recent output. For a clearer fading effect, stop "playing" the input .wav file temporarily.

Experiment with the three sliders for different echo effects.

```
//Echo_control.c Echo using two interrupts for read and write
//3 sliders to control effects: buffer size, amplitude, fading

short input, output;
short bufferlength = 1000;            //initial buffer size
short i = 0;                          //buffer index
short buffer[8000];                   //max size of buffer
short delay = 1;                      //determines size of buffer
short delay_flag = 1;                 //flag if buffer size changes
short amplitude = 1;                  //amplitude control by slider
short echo_type = 0;                  //no fading (1 for fading)

interrupt void c_int11()              //ISR INT11 to write
{
 short new_count;                     //count for new buffer

 output=input+0.1*amplitude*buffer[i]; //newest+oldest sample
 if (echo_type == 1)                  //if fading is desired
  {
   new_count = (i-1) % bufferlength;  //previous buffer location
   buffer[new_count] = output;        //to store most recent output
  }
 output_sample(output);               //output delayed sample
}

interrupt void c_int12()              //ISR INT12 to read
{
 input = input_sample();              //newest input sample data
 if (delay_flag != delay)             //if delay has changed
  {                                   //->new buffer size
   delay_flag = delay;                //reint for future change
   bufferlength = 1000*delay;         //new buffer length
   i = 0;                             //reinit buffer count
  }
 buffer[i] = input;                   //store input sample
 i++;                                 //increment buffer index
 if (i == bufferlength) i=0;          //if @ end of buffer reinit
}

main()
{
 comm_intr();                         //init DSK, codec, McBSP
 while(1);                            //infinite loop
}
```

FIGURE 2.18. Echo generation with controls for different effects (echo_control.c).

```
//Echo_control.gel Sliders vary time delay, amplitude, and type of echo

menuitem "Echo Control"

slider Amplitude(1,8,1,1,amplitude_parameter)      /*incr by 1, up to 8*/
  {
  amplitude = amplitude_parameter;                 /*vary amplit of echo*/
  }
slider Delay(1,8,1,1,delay_parameter)              /*incr by 1, up to 8*/
  {
  delay = delay_parameter;                         /*vary delay of echo*/
  }
slider Type(0,1,1,1,echo_typeparameter)            /*incr by 1, up to 1*/
  {
  echo_type = echo_typeparameter;                  /*echo type for fading*/
  }
```

FIGURE 2.19. GEL file for echo control of amplitude, delay, and fading (echo_control.gel).

Example 2.12: Sine Generation with Table Values Generated within Program (sinegen_table)

This example creates one period of sine data values for a table. Then these values are output for generating a sine wave. Figure 2.20 shows a listing of the program sinegen_table.c, which implements this project. The frequency generated is $f = F_s/$(number of points) $= 8000/10 = 800\,Hz$.

This project, which uses the transmit interrupt INT11 should be build and run as **sinegen_table**. Verify a sine wave generated with a frequency of 800 Hz. Change the number of points to generate a 400-Hz sine wave (only table_size needs to be changed).

Example 2.13: Sine Generation with Table Created by MATLAB (sin1500MATL)

This example illustrates the generation of a sinusoid using a lookup table created with MATLAB. Figure 2.21 shows a listing of the MATLAB program sin1500.m, which generates a file with 128 data points with 24 cycles. The sine-wave frequency generated is

$$f = F_s \,(\text{number of cycles})/(\text{number of points}) = 1500\,Hz$$

Run sin1500.m within MATLAB and verify the header file sin1500.h with 128 points, as shown in Figure 2.22. Different numbers of points representing sinusoidal

```
//Sinegen_table.c Generates a sinusoid for a look-up table

#include <math.h>
#define table_size (short)10      //set table size
short sine_table[table_size];     //sine table array
short i;

interrupt void c_int11()          //interrupt service routine
{
 output_sample(sine_table[i]);    //output each sine value
 if (i < table_size - 1) ++i;     //incr index until end of table
    else i = 0;                   //reinit index if end of table
 return;                          //return from interrupt
}

void main()
{
float pi=3.14159;

for(i = 0; i < table_size; i++)
  sine_table[i]=10000*sin(2.0*pi*i/table_size); //scaled values

i = 0;
comm_intr();                      //init DSK, codec, McBSP
while(1);                         //infinite loop
}
```

FIGURE 2.20. Sine-wave generation program using table generated within program (sinegen_table.c).

```
%sin1500.m Generates 128 points representing sin(1500) Hz
%Creates file sin1500.h
for i=1:128
  sine(i) = round(1000*sin(2*pi*(i-1)*1500/8000)); %sin(1500)
end

fid = fopen('sin1500.h','w');            %open/create file
fprintf(fid,'short sin1500[128]={');     %print array name,"={"
fprintf(fid,'%d, ' ,sine(1:127));        %print 127 points
fprintf(fid,'%d' ,sine(128));            %print 128th point
fprintf(fid,'};\n');                     %print closing bracket
fclose(fid);                             %close file
```

FIGURE 2.21. MATLAB program to generate a lookup table for sine-wave data (sin1500.m).

```
short sin1500[128]={0, 924, 707, -383, -1000, -383, 707, 924, 0,
-924, -707, 383, 1000, 383, -707, -924, 0, 924, 707, -383,
-1000, -383, 707, 924, 0, -924, -707, 383, 1000, 383, -707,
-924, 0, 924, 707, -383, -1000, -383, 707, 924, 0, -924, -707,
383, 1000, 383, -707, -924, 0, 924, 707, -383, -1000, -383, 707,
924, 0, -924, -707, 383, 1000, 383, -707, -924, 0, 924, 707,
-383, -1000, -383, 707, 924, 0, -924, -707, 383, 1000, 383,
-707, -924, 0, 924, 707, -383, -1000, -383, 707, 924, 0, -924,
-707, 383, 1000, 383, -707, -924, 0, 924, 707, -383, -1000,
-383, 707, 924, 0, -924, -707, 383, 1000, 383, -707, -924, 0,
924, 707, -383, -1000, -383, 707, 924, 0, -924, -707, 383, 1000,
383, -707, -924};
```

FIGURE 2.22. Sine table-lookup header file generated by MATLAB (sin1500.h).

```
//Sin1500MATL.c Generates sine from table created with MATLAB

#include "sin1500.h"         //sin(1500) created with MATLAB
short i=0;

interrupt void c_int11()
{
 output_sample(sin1500[i]);  //output each sine value
 if (i < 127) ++i;           //incr index until end of table
   else i = 0;
 return;                     //return from interrupt
}

void main()
{
 comm_intr();                //init DSK, codec, McBSP
 while(1);                   //infinite loop
}
```

FIGURE 2.23. Sine generation program using header file with sine data values generated with MATLAB (sin1500MATL.c).

signals of different frequencies can readily be obtained with minor changes in the MATLAB program *sin1500.m.*

Figure 2.23 shows a listing of the C source file *sin1500MATL.c*, which implements this project in real time. This program includes the header file generated by MATLAB. See also Example 2.12, which generates the table within the main C source program in lieu of using MATLAB.

Build and run this project as **sin1500MATL**. Verify that the output is a 1500-Hz sine-wave signal. Within CCS, be careful when you view the header file sin1500.h so as not to truncate it.

```
//AM.c AM using table for carrier and baseband signals

short amp = 1;

void main()
{
 short baseband[20]={1000,951,809,587,309,0,-309,-587,-809,-951,
        -1000,-951,-809,-587,-309,0,309,587,809,951}; //400-Hz baseband
 short carrier[20] ={1000,0,-1000,0,1000,0,-1000,0,1000,0,
        -1000,0,1000,0,-1000,0,1000,0,-1000,0}; //2-kHz carrier
 short output[20];
 short k;

 comm_poll();                               //init DSK, codec, McBSP
 while(1)                                   //infinite loop
  {
   for (k=0; k<20; k++)
    {
     output[k]= carrier[k] + ((amp*baseband[k]*carrier[k]/10)>>12);
     output_sample(20*output[k]);           //scale output
    }
  }
}
```

FIGURE 2.24. Amplitude modulation program (AM.c).

Example 2.14: Amplitude Modulation (AM)

This example illustrates an amplitude modulation (AM) scheme. Figure 2.24 shows a listing of the program AM.c, which generates an AM signal. The buffer *baseband* contains 20 points and represents a baseband cosine signal with a frequency of $f = F_s/20 = 400$ Hz. The buffer *carrier* also contains 20 points and represents a carrier signal with a frequency of $f = F_s$ (number of cycles)/(number of points) = F_s/(number points per cycle) = 2 kHz. The output equation shows the baseband signal being modulated by the carrier signal. The variable amp is used to vary the modulation. The C source program AM.c is not interrupt-driven. Choose the appropriate vector support file.

Build and implement this project as **AM**. Verify that the output consists of the 2-kHz carrier signal and two sideband signals. The sideband signals are at the frequency of the carrier signal + or − the frequency of the sideband signal, or at 1600 and 2400 Hz.

Load the GEL file AM.gel, increase the variable amp, and verify the baseband signal being modulated. Note that the product of the carrier and baseband signals (within the output equation) is scaled by 2^{12} (shifted right by 12). The voice scrambler (Example 4.9) makes further use of modulation in order to scramble an input signal.

*Alternative AM with External Input for Sideband (**AM_extin**)*
The program *AM_extin.c* (on the accompanying disk) illustrates an alternative modulating scheme to obtain an AM signal using an external input as the sideband signal and a 2-kHz carrier signal from a lookup table.

   Build this project as **AM_extin**. Test this project using a sinusoidal sideband signal with an amplitude below 0.35 V and a frequency less than 2 kHz. Such a small external input signal yields a more stable output. Note that a frequency of more than 2 kHz will cause aliasing.

Example 2.15: Sweep Sinusoid Using Table with 8000 Points (*sweep8000*)

Figure 2.25 shows a listing of the program *sweep8000.c*, which generates a sweeping sinusoidal signal using a table lookup with 8000 points. The header file *sine8000_table.h* contains the 8000 data points that represent a one-cycle sine

```
//Sweep8000.c Sweep sinusoid using table with 8000 points

#include "sine8000_table.h"          //one cycle with 8000 points
short start_freq = 100;              //initial frequency
short stop_freq = 3500;             //maximum frequency
short step_freq = 200;              //increment/step frequency
short amp = 30;                     //amplitude
short delay_msecs = 1000;          //# of msec at each frequency
short freq;
short t;
short i = 0;

void main()
{
 comm_poll();                       //init DSK, codec, McBSP
 while(1)                           //infinite loop
  {
   for(freq=start_freq; freq<=stop_freq; freq+=step_freq)
    {                               //step thru freqs
     for(t=0; t<8*delay_msecs; t++) //output 8*delay_msecs samples
      {                             // at each freq
       output_sample(amp*sine8000[i]); //output
       i = (i + freq) % 8000;       //next sample is + freq in table
      }
    }
  }
}
```

FIGURE 2.25. Program to generate sweeping sinusoid using table lookup with 8000 points (sweep8000.c).

wave. Since the output rate is $F_s = 8\,\text{kHz}$, 8000 points are chosen to represent a 1-second interval. The file *sine8000_table.h* (on the disk) is generated with MATLAB using

1000*sin(2*pi*i*start_freq/8000)

Figure 2.26 shows a partial listing of the file *sine8000_table.h*.

The initial frequency is set at 100 Hz and increments every 200 Hz until a stop frequency of 3500 Hz is reached. The frequencies generated are 100, 300, 500, ..., 3500 Hz, and each frequency is generated for 1 second.

Increase `delay_msecs` from 1000 to 2000 for a slower sweep, since each frequency would be generated for 2 seconds. If `step_freq` is increased to 700, the frequencies generated would be 100, 800, 1500, 2200, and 2900 Hz.

The index `i` is incremented by `i + freq`, which determines the values chosen from the table (see also Example 2.4). For example, to generate 100 Hz, every 100th value in the table is selected to output 80 data points, corresponding to 1 cycle, 8000

//sine8000_table.h Sine table with 8000 points generated with MATLAB

```
short  sine8000[8000]=
{0,  1,  2,  2,  3,  4,  5,  5,
6,  7,  8,  9,  9,  10,  11,  12,
13,  13,  14,  15,  16,  16,  17,  18,
19,  20,  20,  21,  22,  23,  24,  24,
25,  26,  27,  27,  28,  29,  30,  31,
31,  32,  33,  34,  35,  35,  36,  37,
38,  38,  39,  40,  41,  42,  42,  43,
44,  45,  46,  46,  47,  48,  49,  49,
50,  51,  52,  53,  53,  54,  55,  56,
57,  57,  58,  59,  60,  60,  61,  62,
63,  64,  64,  65,  66,  67,  67,  68,
69,  70,  71,  71,  72,  73,  74,  75,
75,  76,  77,  78,  78,  79,  80,  81,
82,  82,  83,  84,  85,  86,  86,  87,
88,  89,  89,  90,  91,  92,  93,  93,
94,  95,  96,  96,  97,  98,  99,  100,
100,  101,  102,  103,  103,  104,  105,  106,
107,  107,  108,  109,  110,  111,  111,  112,
   .
   .
   .
-13,  -12,  -11,  -10,  -9,  -9,  -8,  -7,
-6,  -5,  -5,  -4,  -3,  -2,  -2,  -1};
```

FIGURE 2.26. Partial listing of sine with 8000 data points (`sine8000_table.h`).

```
//Noise_gen.c Pseudo-random sequence generation

#include "noise_gen.h"              //header file for noise sequence
int fb;
shift_reg sreg;                     //shift reg structure

interrupt void c_int11()            //interrupt service routine
{
 int prnseq;                        //for pseudo-random sequence

 if(sreg.bt.b0)                     //sequence{1,-1}based on bit b0
     prnseq = -8000;                //scaled negative noise level
 else
     prnseq = 8000;                 //scaled positive noise level
 fb =(sreg.bt.b0)^(sreg.bt.b1);     //XOR bits 0,1
 fb ^=(sreg.bt.b11)^(sreg.bt.b13);  //with bits 11,13 ->fb
 sreg.regval<<=1;                   //shift register 1 bit to left
 sreg.bt.b0 = fb;                   //close feedback path

 output_sample(prnseq);             //output scaled sequence
 return;                            //return from interrupt
}

void main()
{
 sreg.regval = 0xFFFF;              //set shift register
 fb = 1;                            //initial feedback value
 comm_intr();                       //init DSK, codec, McBSP
 while (1);                         //infinite loop
}
```

FIGURE 2.27. Pseudorandom noise sequence generation program (noise_gen.c).

points over 100 cycles. With this scheme, 8000 points are always used to generate each frequency over x cycles per second.

Build and run this project as **sweep8000**. Verify the output as a sweeping sinusoid. Note that the source program *sweep8000.c* is not interrupt-driven (use the appropriate vector file). A slider can be used to control the amplitude of the frequency generated with the variable *amp*, the duration at each frequency with *delay_msecs* (sweep speed), and the incremental frequency with *step_freq*.

Example 2.16: Pseudorandom Noise Sequence Generation (*noise_gen*)

The program noise_gen.c, shown in Figure 2.27, generates a pseudorandom noise sequence. It uses a software-based implementation of a maximal-length sequence technique for generating a pseudorandom sequence. An initial 16-bit seed is assigned to a register. Bits b0, b1, b11, and b13 are XORed and the result placed into a feedback variable. The register with the initial seed value is then shifted one bit to the left. The feedback variable is then assigned to bit b0 of the register. A scaled minimum or maximum is assigned to `prnseq`, depending on whether the register's bit b0 is zero or 1. This scaled value corresponds to the noise-level amplitude. The header file `noise_gen.h` (on the disk) defines the shift register bits.

Build and run this project as **noise_gen**. You can view the noise in the time domain or hear it. Increase the noise-level amplitude for a scaled value of ±16,000 and verify that the noise generated is louder. Connect the output to a spectrum analyzer. Verify that the output spectrum is relatively flat until the cutoff frequency of approximately 3500 Hz, which represents the bandwidth of the antialiasing filter on the codec.

REFERENCES

1. *TLC320AD535C/I Data Manual Dual Channel Voice/Data Codec*, SLAS202A, Texas Instruments, Dallas, TX, 1999.

2. S. Norsworthy, R. Schreier, and G. Temes, *Delta–Sigma Data Converters: Theory, Design and Simulation*, IEEE Press, Piscataway, NJ, 1997.

3. P. M. Aziz, H. V. Sorensen, and J. Van Der Spiegel, An overview of sigma delta converters, *IEEE Signal Processing*, Jan. 1996.

4. J. C. Candy and G. C. Temes, eds., *Oversampling Delta–Sigma Data Converters: Theory, Design and Simulation*, IEEE Press, Piscataway, NJ, 1992.

5. C. W. Solomon, Switched-capacitor filters, *IEEE Spectrum*, June 1988.

6. *PCM3002/PCM3003 16-/20-Bit Single-Ended Analog Input/Output Stereo Audio Codecs*, SBAS079, Texas Instruments, Dallas, TX, 2000.

7. *TMS320C6000 McBSP: AC'97 Codec Interface*, SPRA528, Texas Instruments, Dallas, TX, 1999.

3

Architecture and Instruction Set of the C6x Processor

- Architecture and instruction set of the TMS320C6x processor
- Addressing modes
- Assembler directives
- Linear assembler
- Programming examples using C, assembly, and linear assembly code

3.1 INTRODUCTION

Texas Instruments introduced the first-generation TMS32010 digital signal processor in 1982, the TMS320C25 in 1986 [1], and the TMS320C50 in 1991. Several versions of each of these processors—C1x, C2x, and C5x—are available with different features, such as faster execution speed. These 16-bit processors are all fixed-point processors and are code-compatible.

In a von Neumann architecture, program instructions and data are stored in a single memory space. A processor with a von Neumann architecture can make a read or a write to memory during each instruction cycle. Typical DSP applications require several accesses to memory within one instruction cycle. The fixed-point processors C1x, C2x, and C5x are based on a modified Harvard architecture with separate memory spaces for data and instructions that allow concurrent accesses.

Quantization error or round-off noise from an ADC is a concern with a fixed-point processor. An ADC uses only a best-estimate digital value to represent an input. For example, consider an ADC with a word length of 8 bits and an input range of $\pm 1.5\,V$. The steps represented by the ADC are: input range/$2^8 = 3/256 = 11.72\,mV$. This produces errors which can be up to $\pm(11.72\,mV)/2 = \pm5.86\,mV$. Only a best estimate can be used by the ADC to represent input values that are not multiples of

11.72 mV. With an 8-bit ADC, 2^8 or 256 different levels can represent the input signal. An ADC with a larger word length such as a 16-bit ADC (currently very common) can reduce the quantization error, yielding a higher resolution. The more bits that an ADC has, the better it can represent an input signal.

The TMS320C30 floating-point processor was introduced in the late 1980s. The C31, C32, and the more recent C33 are all members of the C3x family of floating-point processors [2,3]. The C4x floating-point processors, introduced subsequently, are code-compatible with the C3x processors and are based on the modified Harvard architecture [4].

The TMS320C6201 (C62x), announced in 1997, is the first member of the C6x family of fixed-point digital signal processors. Unlike the previous fixed-point processors, C1x, C2x, and C5x, the C62x is based on a very-long-instruction-word (VLIW) architecture, still using separate memory spaces for instructions and data as with the Harvard architecture. The VLIW architecture has simpler instructions, but more are needed for a task than with a conventional DSP architecture.

The C62x is not code-compatible with the previous generation of fixed-point processors. Subsequently, the TMS320C6701 (C67x) floating-point processor was introduced as another member of the C6x family of processors. The instruction set of the C62x fixed-point processor is a subset of the instruction set of the C67x processor. Appendix A contains a list of instructions available on the C6x processors. A recent addition to the family of the C6x processors is the fixed-point C64x.

An application-specific integrated circuit (ASIC) has a DSP core with customized circuitry for a specific application. A C6x processor can be used as a standard general-purpose digital signal processor programmed for a specific application. Specific-purpose digital signal processors are the modem, echo canceler, and others.

A fixed-point processor is better for devices that use batteries, such as cellular phones, since it uses less power than does an equivalent floating-point processor. The fixed-point processors, C1x, C2x, and C5x are 16-bit processors with limited dynamic range and precision. The C6x fixed-point processor is a 32-bit processor with improved dynamic range and precision. In a fixed-point processor, it is necessary to scale the data. Overflow, which occurs when an operation such as the addition of two numbers produces a result with more bits than can fit within a processor's register, becomes a concern.

A floating-point processor is generally more expensive since it has more "real estate" or is a larger chip because of additional circuitry necessary to handle integer as well as floating-point arithmetic. Several factors, such as cost, power consumption, and speed, come into play when choosing a specific digital signal processor. The C6x processors are particularly useful for applications requiring intensive computations. Family members of the C6x include both fixed-point (e.g., C62x, C64x) and floating-point processors (e.g., C67x). Other digital signal processors are also available, from companies such as Motorola and Analog Devices [5].

Other architectures include the Super Scalar, which requires special hardware to determine which instructions are executed in parallel. The burden is then on the processor more than on the programmer as in the VLIW architecture. It does not execute necessarily the same group of instructions, and as a result, it is difficult to time. Thus, it is rarely used in DSP.

3.2 TMS320C6x ARCHITECTURE

The TMS320C6711 onboard the DSK is a floating-point processor based on the VLIW architecture [6–9]. Internal memory includes a two-level cache architecture with 4 kB of level 1 program cache (L1P), 4 kB of level 1 data cache (L1D), and 64 kB of RAM or level 2 cache for data/program allocation (L2). It has a glueless (direct) interface to both synchronous memories (SDRAM and SBSRAM) and asynchronous memories (SRAM and EPROM). Synchronous memory requires clocking but provides a compromise between static SRAM and dynamic SDRAM, with SRAM being faster but more expensive than DRAM.

On-chip peripherals include two multichannel buffered serial ports (McBSPs), two timers, a 16-bit host port interface (HPI), and a 32-bit external memory interface (EMIF). It requires 3.3 V for I/O and 1.8 V for the core (internal). Internal buses include a 32-bit program address bus, a 256-bit program data bus to accommodate eight 32-bit instructions, two 32-bit data address buses, two 64-bit data buses, and two 64-bit store data buses. With a 32-bit address bus, the total memory space is $2^{32} = 4$ GB, including four external memory spaces: CE0, CE1, CE2, and CE3. Figure 3.1 shows a functional block diagram of the C6711 processor included with CCS.

Independent memory banks on the C6x allow for two memory accesses within one instruction cycle. Two independent memory banks can be accessed using two

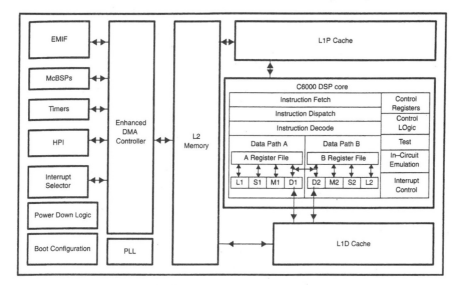

FIGURE 3.1. Functional block diagram of TMS320C6x (Courtesy of Texas Instruments).

independent buses. Since internal memory is organized into memory banks, two loads or two stores instructions can be performed in parallel. No conflict results if the data accessed are in different memory banks. Separate buses for program, data, and direct memory access (DMA) allow the C6x to perform concurrent program fetches, data read and write, and DMA operations. With data and instructions residing in separate memory spaces, concurrent memory accesses are possible. The C6x has a byte-addressable memory space. Internal memory is organized as separate program and data memory spaces, with two 32-bit internal ports (two 64-bit ports with the C64x) to access internal memory.

The C6711 on the DSK includes 72 kB of internal memory, which starts at 0x00000000, and 16 MB of external SDRAM, mapped through CE0 starting at 0x80000000. The DSK also includes 128 kB of Flash memory onboard, starting at 0x90000000. A two-level internal memory block diagram is shown in Figure 3.2, included with CCS [7]. Table 3.1 shows the memory map. A schematic diagram of the DSK is included with CCS (*C6711dsk_schematics.pdf*).

With a clock of 150 MHz onboard the DSK, one can ideally achieve two multiplies and accumulates per cycle, for a total of 300 million multiplies and accumu-

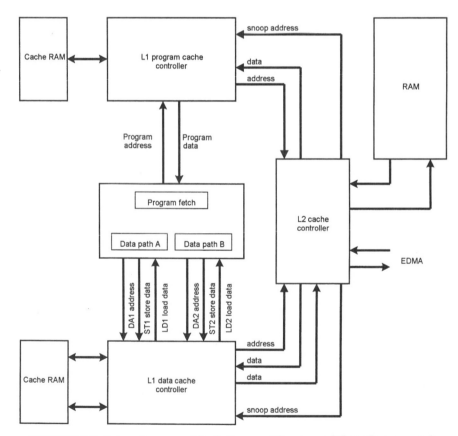

FIGURE 3.2. Internal memory block diagram (Courtesy of Texas Instruments).

TABLE 3.1   Memory Map Summary

Address Range (Hex)	Size (Bytes)	Description of Memory Block
0000 0000—0000 FFFF	64K	Internal RAM (L2)
0001 0000—017F FFFF	24M–64K	Reserved
0180 0000—0183 FFFF	256K	Internal configuration bus EMIF registers
0184 0000—0187 FFFF	256K	Internal configuration bus L2 control registers
0188 0000—018B FFFF	256K	Internal configuration bus HPI register
018C 0000—018F FFFF	256K	Internal configuration bus McBSP 0 registers
0190 0000—0193 FFFF	256K	Internal configuration bus McBSP 1 registers
0194 0000—0197 FFFF	256K	Internal configuration bus timer 0 registers
0198 0000—019B FFFF	256K	Internal configuration bus timer 1 registers
019C 0000—019F FFFF	256K	Internal configuration bus interrupt selector registers
01A0 0000—01A3 FFFF	256K	Internal configuration bus EDMA RAM and registers
01A4 0000—01FF FFFF	6M–256K	Reserved
0200 0000—0200 0033	52	QDMA registers
0200 0034—2FFF FFFF	736M–52	Reserved
3000 0000—3FFF FFFF	256M	McBSP 0/1 data
4000 0000—7FFF FFFF	1G	Reserved
8000 0000—8FFF FFFF	256M	External memory interface CE0
9000 0000—9FFF FFFF	256M	External memory interface CE1
A000 0000—AFFF FFFF	256M	External memory interface CE2
B000 000—BFFF FFFF	256M	External memory interface CE3
C000 0000—FFFF FFFF	1G	Reserved

Source: Courtesy of Texas Instruments [7].

lates (MACs) per second. With six of the eight functional units in Figure 3.1 (not the .D units described below) capable of handling floating-point operations, it is possible to perform 900 million floating-point operations per second (MFLOPS). Operating at 150 MHz, this translates to 1200 million instructions per second (MIPS) with a 6.67-ns instruction cycle time.

3.3 FUNCTIONAL UNITS

The CPU consists of eight independent functional units divided into two data paths A and B, as shown in Figure 3.1. Each path has a unit for multiply operations (.M), for logical and arithmetic operations (.L), for branch, bit manipulation, and arithmetic operations (.S), and for loading/storing and arithmetic operations (.D). The .S and .L units are for arithmetic, logical, and branch instructions. All data transfers make use of the .D units.

The arithmetic operations, such as subtract or add (SUB or ADD), can be performed by all the units except the .M units (one from each data path). The eight functional units consist of four floating/fixed-point ALUs (two .L and two .S), two fixed-point ALUs (.D units), and two floating/fixed-point multipliers (.M units). Each functional unit can read directly from or write directly to the register file within its own path. Each path includes a set of sixteen 32-bit registers, A0 through A15 and B0 through B15. Units ending in 1 write to register file A, and units ending in 2 write to register file B.

Two cross-paths (1x and 2x) allow functional units from one data path to access a 32-bit operand from the register file on the opposite side. There can be a maximum of two cross-path source reads per cycle. Each functional unit side can access data from the registers on the opposite side using a cross-path (i.e., the functional units on one side can access the register set from the other side). There are 32 general-purpose registers, but some of them are reserved for specific addressing or are used for conditional instructions.

3.4 FETCH AND EXECUTE PACKETS

The architecture VELOCITI, introduced by TI, is derived from the VLIW architecture. An execute packet (EP) consists of a group of instructions that can be executed in parallel within the same cycle time. The number of EPs within a fetch packet (FP) can vary from one (with eight parallel instructions) to eight (with no parallel instructions). The VLIW architecture was modified to allow more than one EP to be included within an EP.

The least significant bit of every 32-bit instruction is used to determine if the next or subsequent instruction belongs in the same EP (if 1) or is part of the next EP (if 0). Consider an FP with three EPs: EP1, with two parallel instructions, and EP2 and EP3, each with three parallel instructions, as follows:

```
        Instruction A
||      Instruction B

        Instruction C
||      Instruction D
||      Instruction E

        Instruction F
||      Instruction G
||      Instruction H
```

EP1 contains the two parallel instructions A and B; EP2 contains the three parallel instructions C, D, and E; and EP3 contains the three parallel instructions F, G, and H. The FP would be as shown in Figure 3.3. Bit 0 (LSB) of each 32-bit instruction contains a "p" bit that signals whether it is in parallel with a subsequent instruction. For example, the "p" bit of instruction B is zero, denoting that it is not within the same EP as the subsequent instruction C. Similarly, instruction E is not within the same EP as instruction F.

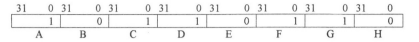

FIGURE 3.3. One fetch packet with three execute packets, showing the "p" bit of each instruction.

3.5 PIPELINING

Pipelining is a key feature in a digital signal processor to get parallel instructions working properly, requiring careful timing. There are three stages of pipelining: program fetch, decode, and execute.

1. The *program fetch stage* is composed of four phases:
   (a) *PG*: program address generate (in the CPU) to fetch an address
   (b) *PS*: program address send (to memory) to send the address
   (c) *PW*: program address ready wait (memory read) to wait for data
   (d) *PR*: program fetch packet receive (at the CPU) to read opcode from memory
2. The *decode stage* is composed of two phases:
   (a) *DP*: to dispatch all the instructions within an FP to the appropriate functional units
   (b) *DC*: instruction decode
3. The *execute stage* is composed of from six phases (with fixed point) to 10 phases (with floating point), due to delays (latencies) associated with the following instructions:
   (a) Multiply instruction, which consists of two phases due to one delay
   (b) Load instruction, which consists of five phases due to four delays
   (c) Branch instruction, which consists of six phases due to five delays

Table 3.2 shows the pipeline phases, and Table 3.3 shows the pipelining effects. The first row in Table 3.3 represents cycle 1, 2, . . . , 12. Each subsequent row represents an FP. The rows represented PG, PS, . . . , illustrate the phases associated with each FP. The program generate (PG) of the first FP starts in cycle 1, and the PG of the second FP starts in cycle 2, and so on. Each FP takes four phases for program fetch and two phases for decoding. However, the execution phase can take from 1 to 10 phases (not all execution phases are shown in Table 3.3). We are assuming that each FP contains one execute packet (EP).

For example, at cycle 7, while the instructions in the first FP are in the first execution phase E1 (which may be the only one), the instructions in the second FP are in the decoding phase, the instructions in the third FP are in the dispatching phase, and so on. All seven instructions are proceeding through the various phases. Therefore, at cycle 7, "the pipeline is full."

TABLE 3.2   Pipeline Phases

Program Fetch				Decode		Execute
PG	PS	PW	PR	DP	DC	E1–E6 (E1–E10 for double precision)

TABLE 3.3   Pipelining Effects

					Clock Cycle						
1	2	3	4	5	6	7	8	9	10	11	12
PG	PS	PW	PR	DP	DC	E1	E2	E3	E4	E5	E6
	PG	PS	PW	PR	DP	DC	E1	E2	E3	E4	E5
		PG	PS	PW	PR	DP	DC	E1	E2	E3	E4
			PG	PS	PW	PR	DP	DC	E1	E2	E3
				PG	PS	PW	PR	DP	DC	E1	E2
					PG	PS	PW	PR	DP	DC	E1
						PG	PS	PW	PR	DP	DC

Most instructions have one execute phase. Instructions such as multiply (MPY), load (LDH/LDW), and branch (B) take two, five, and six phases, respectively. Additional execute phases are associated with floating-point and double-precision types of instructions, which can take up to 10 phases. For example, the double-precision multiply operation (MPYDP), available on the C67x, has nine delay slots, so that the execution phase takes a total of 10 phases.

The *functional unit latency*, which represents the number of cycles that an instruction ties up a functional unit, is 1 for all instructions except double-precision instructions, available with the floating-point C67x. Functional unit latency is different from a delay slot. For example, the instruction MPYDP has four functional unit latencies but nine delay slots. This implies that no other instruction can use the associated multiply functional unit for four cycles. A store has no delay slot but finishes its execution in the third execution phase of the pipeline.

If the outcome of a multiply instruction such as MPY is used by a subsequent instruction, a NOP (no operation) must be inserted after the MPY instruction for the pipelining to operate properly. Four or five NOPs are to be inserted in case an instruction uses the outcome of a load or a branch instruction, respectively.

3.6 REGISTERS

Two sets of register files, each set with 16 registers, are available: register file A (A0 through A15) and register file B (B0 through B15). Registers A0, A1, B0, B1, and B2 are used as conditional registers. Registers A4 through A7 and B4 through B7 are used for circular addressing. Registers A0 through A9 and B0 through B9 (except B3) are temporary registers. Any of the registers A10 through A15 and

B10 through B15 used are saved and later restored before returning from a subroutine.

A 40-bit data value can be contained across a register pair. The 32 least significant bits (LSBs) are stored in the even register (e.g., A2) and the remaining 8 bits are stored in the 8 LSBs of the next-upper (odd) register (A3). A similar scheme is used to hold a 64-bit double-precision value within a pair of registers (even and odd).

These 32 registers are considered as general-purpose registers. Several special-purpose registers are also available for control and interrupts: for example, the address mode register (AMR) used for circular addressing and interrupt control registers, as shown in Appendix B.

3.7 LINEAR AND CIRCULAR ADDRESSING MODES

Addressing modes determine how one accesses memory. They specify how data are accessed, such as retrieving an operand indirectly from a memory location. Both linear and circular modes of addressing are supported. The most commonly used mode is the indirect addressing of memory.

3.7.1 Indirect Addressing

Indirect addressing can be used with or without displacement. Register R represents one of the 32 registers A0 through A15 and B0 through B15 that can specify or point to memory addresses. As such, these registers are pointers. Indirect addressing mode uses a "*" in conjunction with one of the 32 registers. To illustrate, consider R as an address register.

1. *R*. Register R contains the address of a memory location where a data value is stored.
2. *R++(d)*. Register R contains the memory address (location). After the memory address is used, R is postincremented (modified), such that the new address is the current address offset by the displacement value d. If d = 1 (by default), the new address is R + 1, or R is incremented to the next-higher address in memory. A double minus (––) instead of a double plus would update or postdecrement the address to R – d.
3. *++R(d)*. The address is preincremented or offset by d, such that the current address is R + d. A double minus would predecrement the memory address so that the current address is R – d.
4. *+R(d)*. The address is preincremented by d, such that the current address is R + d (as with the preceding case). However, in this case, R preincrements without modification. Unlike the previous case, R is not updated or modified.

3.7.2 Circular Addressing

Circular addressing is used to create a circular buffer. This buffer is created in hardware and is very useful in several DSP algorithms, such as in digital filtering or correlation algorithms where data need to be updated. An example in Chapter 4 illustrates the implementation of a digital filter using a circular buffer to update the "delay" samples.

The C6x has dedicated hardware to allow a circular type of addressing. This addressing mode can be used in conjunction with a circular buffer to update samples by shifting data without the overhead created by shifting data directly. As a pointer reaches the end or "bottom" location of a circular buffer that contains the last element in the buffer, and is then incremented, the pointer is automatically wrapped around or points to the beginning or "top" location of the buffer that contains the first element.

Two independent circular buffers are available using BK0 and BK1 within the address mode register (AMR), as shown in Appendix B. The eight registers A4 through A7 and B4 through B7, in conjunction with the two .D units, can be used as pointers (all registers can be used for linear addressing). The following code segment illustrates the use of a circular buffer using register B2 (only side B can be used) to set the appropriate values within AMR:

```
MVK    .S2   0x0004,B2   ;lower 16 bits to B2. Select A5 as pointer
MVKLH  .S2   0x0005,B2   ;upper 16 bits to B2. Select B0, set N = 5
MVC    .S2   B2,AMR      ;move 32 bits of B2 to AMR
```

The two move instructions MVK and MVKLH (using the .S unit) move 0x0004 into the 16 LSBs of register B2 and 0x0005 into the 16 MSBs of B2. The MVC (move constant) instruction is the only instruction that can access the AMR and the other control registers (shown in Appendix B) and executes only on the B side in conjunction with the functional units and registers on the side B. A 32-bit value is created in B2, which is then transferred to AMR with the instruction MVC to access AMR [6].

The value 0x0004 = $(0100)_b$ into the 16 LSBs of AMR sets bit 2 (third bit) to 1 and all other bits to zero. This sets the mode to 01 and selects register A5 as the pointer to a circular buffer using block BK0.

Table 3.4 shows the modes associated with registers A4 through A7 and B4 through B7. The value 0x0005 = $(0101)_b$ into the 16 MSBs of AMR sets bits 16 and 18 to 1 (other bits to zero). This corresponds to the value of N used to select the size of the buffer as $2^{N+1} = 64$ bytes using BK0. For example, if a buffer size of 128 is desired using BK0, the upper 16 bits of AMR are set to $(0110)_b$ = 0x0006. If assembly code is used for the circular buffer, as execution returns to a calling C function, AMR needs to be reinitialized to the default linear mode. Hence the pointer's address must be saved.

TABLE 3.4   AMR Mode and Description

Mode	Description
0 0	For linear addressing (default on reset)
0 1	For circular addressing using BK0
1 0	For circular addressing using BK1
1 1	Reserved

3.8 TMS320C6x INSTRUCTION SET

3.8.1 Assembly Code Format

An assembly code format is represented by the field

```
Label  ||  [ ]  Instruction  Unit  Operands  ;comments
```

A label, if present, represents a specific address or memory location that contains an instruction or data. The label must be in the first column. The parallel bars (||) are there if the instruction is being executed in parallel with the previous instruction. The subsequent field is optional to make the associated instruction conditional. Five of the registers—A1, A2, B0, B1, and B2—are available to use as conditional registers. For example, [A2] specifies that the associated instruction executes if A2 is not zero. On the other hand, with [!A2], the associated instruction executes if A2 is zero. All C6x instructions can be made conditional with the registers A1, A2, B0, B1, and B2 by determining when the conditional register is zero. The instruction field can be either an assembler directive or a mnemonic. An assembler directive is a command for the assembler. For example,

```
.word value
```

reserves 32 bits in memory and fill with the specified `value`. A mnemonic is an actual instruction that executes at run time. The instruction (mnemonic or assembler directive) cannot start in column 1. The `Unit` field, which can be one of the eight CPU units, is optional. Comments starting in column 1 can begin with either an asterisk or a semicolon, whereas comments starting in any other columns must begin with a semicolon.

Code for the floating-point processors C3x/C4x is not compatible with code for the fixed-point processors C1x, C2x, and C5x/C54x. However, the code for the fixed-point C62x is compatible with the code for the floating-point C67x. C62x code is actually a subset of C67x code. Additional instructions to handle double-precision and floating-point operations are available only on the C67x processor (some additional instructions are also available on the fixed-point C64x processor).

Several code segments are presented to illustrate the C6x instruction set. Assembly code for the C6x processors is very similar to C3x/C4x code. Single-task types of instructions available for the C62x/C67x make it easier to program than either the previous generation of fixed- or floating-point processors. This contributes to an efficient compiler. Additional instructions available on the C64x (but not on the C62x) resemble the multitask types of instructions for C3x/C4x processors. It is very instructive to read the comments in the programs discussed in this book. Appendix B contains a list of the instructions for the C62x/C67x processors.

3.8.2 Types of Instructions

The following illustrates some of the syntax of assembly code. It is optional to specify the eight functional units, although this can be useful during debugging and for code efficiency and optimization, discussed in Chapter 8.

1. *Add/Subtract/Multiply*
   (a) The instruction

```
ADD   .L1   A3,A7,A7   ;add A3 + A7 →A7 (accum in A7)
```

   adds the values in registers A3 and A7 and places the result in register A7. The unit .L1 is optional. If the destination or result is in B7, the unit would be .L2.
   (b) The instruction

```
SUB   .S1   A1,1,A1   ;subtract 1 from A1
```

   subtracts 1 from A1 to decrement it, using the .S unit.
   (c) The parallel instructions

```
     MPY    .M2   A7,B7,B6  ;multiply 16 LSBs of A7,B7 → B6
||   MPYH   .M1   A7,B7,A6  ;multiply 16 MSBs of A7,B7 → A6
```

   multiplies the lower or least significant 16 bits (LSBs) of both A7 and B7 and places the product in B6, in parallel (concurrently within the same execution packet) with a second instruction that multiplies the higher or most significant 16 bits (MSBs) of A7 and B7 and places the result in A6. In this fashion, two multiply/accumulate operations can be executed within a single instruction cycle. This can be used to decompose a sum of products into two sets of sum of products: one set using the lower 16 bits to operate on the first, third, fifth, . . . number, and another set using the higher 16 bits to operate on the second, fourth, sixth, ... number. Note that the parallel symbol is not in column 1.

2. *Load/Store*

(a) The instruction

```
   LDH  .D2  *B2++,B7  ;load (B2) →B7, increment B2
|| LDH  .D1  *A2++,A7  ;load (A2) →A7, increment A2
```

loads into B7 the half-word (16 bits) whose address in memory is specified/pointed by B2. Then register B2 is incremented (postincremented) to point at the next-higher memory address. In parallel is another indirect addressing mode instruction to load into A7 the content in memory, whose address is specified by A2. Then A2 is incremented to point at the next-higher memory address.

The instruction LDW loads a 32-bit word. Two paths using .D1 and .D2 allow for the loading of data from memory to registers A and B using the instruction LDW. The double-word load floating-point instruction LDDW on the C6711 can simultaneously load two 32-bit registers into side A and two 32-bit registers into side B.

(b) The instruction

```
STW  .D2  A1,*+A4[20]  ;store A1→(A4) offset by 20
```

stores the 32-bit word A1 into memory whose address is specified by A4 offset by 20 words (32 bits) or 80 bytes. The adddress register A4 is preincremented with offset, but it is not modified (two plus signs are used if A4 is to be modified).

3. *Branch/Move.* The following code segment illustrates branching and data transfer.

```
Loop MVK  .S1  x,A4       ;move 16 LSBs of x address →A4
     MVKH .S1  x,A4       ;move 16 MSBs of x address →A4
      .
      .
      .
     SUB  .S1  A1,1,A1    ;decrement A1
[A1] B    .S2  Loop       ;branch to Loop if A1 # 0
     NOP       5          ;five no-operation instructions
     STW  .D1  A3,*A7     ;store A3 into (A7)
```

The first instruction moves the lower 16 bits (LSBs) of address *x* into register A4. The second instruction moves the higher 16 bits (MSBs) of address *x* into

A4, which now contains the full 32-bit address of *x*. One must use the instructions MVK/MVKH in order to get a 32-bit constant into a register.

Register A1 is used as a loop counter. After it is decremented with the SUB instruction, it is tested for a conditional branch. Execution branches to the label or address loop if A1 is not zero. If A1 = 0, execution continues and data in register A3 are stored in memory whose address is specified (pointed) by A7.

3.9 ASSEMBLER DIRECTIVES

An assembler directive is a message for the assembler (not the compiler) and is not an instruction. It is resolved during the assembling process and does not occupy memory space as an instruction does. It does not produce executable code. Addresses of different sections can be specified with assembler directives. For example, the assembler directive .sect "my_buffer" defines a section of code or data named my_buffer. The directives .text and .data indicate a section for text and data, respectively. Other assembler directives, such as .ref and .def, are used for undefined and defined symbols, respectively. The assembler creates several sections indicated by directives such as .text for code and .bss for global and static variables.

Other commonly used assembler directives are:

1. .short: to initialize a 16-bit integer.
2. .int: to initialize a 32-bit integer (also .word or .long). The compiler treats a long data value as 40 bits, whereas the C6x assembler treats it as 32 bits.
3. .float: to initialize a 32-bit IEEE single-precision constant.
4. .double: to initialize a 64-bit IEEE double-precision constant.

Initialized values are specified by using the assembler directives .byte, .short, or .int. Unitialized variables are specified using the directive .usect, which creates an uninitialized section (like the .bss section), whereas the directive .sect creates an initialized section. For example, .usect "variable", 128,2 designates an unitialized section named variable, the section size in bytes, and the data alignment in bytes, respectively.

3.10 LINEAR ASSEMBLY

An alternative to C, or assembly code, is linear assembly. An assembler optimizer (in lieu of a C compiler) is used in conjunction with a linear assembly-coded source program (with extension .sa) to create an assembly source program (with extension .asm), in much the same way that a C compiler optimizer is used in conjunction with a C-coded source program. The resulting assembly-coded program produced by the assembler optimizer is typically more efficient than one resulting from the C compiler optimizer. The assembly-coded program resulting from either a C-coded source program or a linear-assembly source program must be assembled to produce an object code.

Linear assembly code programming provides a compromise between coding effort and coding efficiency. The assembler optimizer assigns which functional unit and register to use (optional to be specified by user), finds instructions that can execute in parallel, and performs software pipelining for optimization (discussed in Chapter 8). Two programming examples at the end of this chapter illustrate a C program calling a linear assembly function. Parallel instructions are not valid in a linear assembly program. Specifying the functional unit is optional in a linear assembly program as well as in an assembly program.

Over the last couple of years, the C compiler optimizer has become more and more efficient. Although C code is less efficient (speed performance) than assembly code, it typically involves less coding effort than assembly code, which can be hand-optimized to achieve a 100 percent efficiency but with much greater coding effort.

It may be interesting to note that the C6x assembly code syntax is not as complex as the C2x/C5x or the C3x family of digital signal processors. It is actually simpler to "program" the C6x in assembly. For example, the C3x instruction

```
DBNZD   AR4,LOOP
```

decrements (due to the first D) a loop counter AR4, branches (B) conditionally (if AR4 is nonzero) to the address specified by LOOP, with delay (due to the second D). The branch instruction with delay effectively allows the branch instruction to execute in a single cycle (due to pipelining). Such multitask instructions are not available on the C6x (although recently introduced on the C64x processor). In fact, C6x types of instructions are "simpler." For example, separate instructions are available for decrementing a counter (with a SUB instruction) and branching. The simpler types of instructions are more amenable for a more efficient C compiler.

However, although it is simpler to program in assembly code to perform a desired task, this does not imply or translate to an efficient assembly-coded program. It can be relatively difficult to hand-optimize a program to yield a totally efficient (and meaningful) assembly-coded program.

Linear assembly code is a cross between assembly and C. It uses the syntax of assembly code instructions such as ADD, SUB, and MPY but with operands/registers as used in C. In some cases this provides a good compromise between C and assembly.

Linear assembler directives include

```
.cproc
.endproc
```

to specify a C-callable procedure or section of code to be optimized by the assembler optimizer. Another directive, .reg, is to declare variables and use descriptive names for values that will be stored in registers. Programming examples with C calling an assembly function or C calling a linear assembly function are illustrated later in this chapter.

3.11 ASM STATEMENT WITHIN C

Assembly instructions and directives can be incorporated within a C program using the asm statement. The asm statement can provide access to hardware features that cannot be obtained using C code only. The syntax is

```
asm ("assembly code");
```

The assembly line of code within the set of quotes has the same format as a valid assembly statement. Note that if the instruction has a label, the first character of the label must start after the first quote so that it is in column 1. The assembly statement should be valid since the compiler does not check it for syntax error but copies it directly into the compiled output file. If the assembly statement has a syntax error, the assembler would detect it.

Avoid using asm statements within a C program, especially within a linear assembly program. This is because the assembler optimizer could rearrange lines of code near the asm statements that may cause undesirable results.

3.12 C-CALLABLE ASSEMBLY FUNCTION

Two programming examples are included later in this chapter to illustrate a C program calling an assembly function. Register B3 is preserved and is used to contain the return address of the calling function.

An external declaration of an assembly function called within a C program using extern is optional. For example,

```
extern int func();
```

is optional with the assembly function func returning an integer value.

3.13 TIMERS

Two 32-bit timers can be used to time and count events or to interrupt the CPU. A timer can direct an external ADC to start conversion or the DMA controller to start a data transfer. A timer includes a time period register, which specifies the timer's frequency; a timer counter register, which contains the value of the incrementing counter; and a timer control register, which monitors the timer's status.

3.14 INTERRUPTS

An interrupt can be issued internally or externally. An interrupt stops the current CPU process so that it can perform a required task initiated by the interrupt. The program flow is redirected to an interrupt service routine (ISR). The source of the interrupt can be an ADC, a timer, and so on. Upon an interrupt, the conditions of the current process must be saved so that they can be restored after the interrupt task is performed. On interrupt, registers are saved and processing continues to an ISR. Then the registers are restored.

There are 16 interrupt sources. They include two timer interrupts, four external interrupts, four McBSP interrupts, and four DMA interrupts. Twelve CPU interrupts are available. An interrupt selector is used to choose among the 12 interrupts.

3.14.1 Interrupt Control Registers

The interrupt control registers (Appendix B) follow.

1. **CSR** (control status register): contains the global interrupt enable (GIE) bit and other control/status bits
2. **IER** (interrupt enable register): enables/disables individual interrupts
3. **IFR** (interrupt flag register): displays status of interrupts
4. **ISR** (interrupt set register): sets pending interrupts
5. **ICR** (interrupt clear register): clears pending interrupts
6. **ISTP** (interrupt service table pointer): locates an ISR
7. **IRP** (interrupt return pointer)
8. **NRP** (nonmaskable interrupt return pointer)

Interrupts are prioritized, with Reset having the highest priority. The reset interrupt and nonmaskable interrupt (NMI) are external pins that have the first and second highest priority, respectively. The interrupt enable register (IER) is used to set a specific interrupt and can check if and which interrupt has occurred from the interrupt flag register (IFR).

NMI is nonmaskable, along with Reset. NMI can be masked (disabled) by clearing the NMIE bit within CSR. It is set to zero only upon reset or upon a nonmaskable interrupt. If NMIE is set to zero, all interrupts INT4 through INT15 are disabled. The interrupt registers are shown in Appendix B.

The reset signal is an active-low signal used to halt the CPU, and the NMI signal alerts the CPU to a potential hardware problem. Twelve CPU interrupts with lower priorities are available, corresponding to the maskable signals INT4 through INT15. The priorities of these interrupts are: INT4, INT5, . . . , INT15, with INT4 having the highest priority and INT15 the lowest priority. For a nonmaskable interrupt to occur, the nonmaskable interrupt enable (NMIE) bit must be 1 (active high). On reset (or

TABLE 3.5   Interrupt Service Table

Interrupt	Offset
RESET	000h
NMI	020h
Reserved	040h
Reserved	060h
INT4	080h
INT5	0A0h
INT6	0C0h
INT7	0E0h
INT8	100h
INT9	120h
INT10	140h
INT11	160h
INT12	180h
INT13	1A0h
INT14	1C0h
INT15	1E0h

Source: Courtesy of Texas Instruments.

after a previously set NMI), the NMIE bit is cleared to zero so that a reset interrupt may occur.

To process a maskable interrupt, the global interrupt enable (GIE) bit within the control status register (CSR) and the NMIE bit within the interrupt enable register (IER) are set to 1. GIE is set to 1 with bit 0 of CSR set to 1 and NMIE is set to 1 with bit 1 of IER set to 1. Note that CSR can be ANDed with −2 (using 2's complement, the LSB is zero while all other bits are 1's) to set the GIE bit to zero and disable maskable interrupts globally.

The interrupt enable (IE) bit corresponding to the desirable maskable interrupt is also set to 1. When the interrupt occurs, the corresponding interrupt flag register (IFR) bit is set to 1 to show the interrupt status. To process a maskable interrupt, the following apply:

1. The GIE bit is set to 1.
2. The NMIE bit is set to 1.
3. The appropriate IE bit is set to 1.
4. The corresponding IFR bit is set to 1.

For an interrupt to occur, the CPU must not be executing a delay slot associated with a branch instruction.

The interrupt service table (IST) shown in Table 3.5 is used when an interrupt begins. Within each location is a fetch packet (FP) associated with each interrupt. The table contains 16 FPs, each with eight instructions. The addresses on the right side correspond to an offset associated with each specific interrupt. For example, the FP for interrupt INT11 is at a base address plus an offset of 160h. Since each

TABLE 3.6   Selection of Interrupts Using Interrupt Selector

Interrupt Selector	Type	Description
00000	DSPINT	Host port to DSP interrupt
00001	TINT0	Timer 0 interrupt
00010	TINT1	Timer 1 interrupt
00011	SD_INT	EMIF SDRAM timer interrupt
00100	EXT_INT4	External interrupt pin 4
00101	EXT_INT5	External interrupt pin 5
00110	EXT_INT6	External interrupt pin 6
00111	EXT_INT7	External interrupt pin 7
01000	DMA_INT0	DMA channel 0 interrupt
01001	DMA_INT1	DMA channel 1 interrupt
01010	DMA_INT2	DMA channel 2 interrupt
01011	DMA_INT3	DMA channel 3 interrupt
01100	XINT0	McBSP0 transmit interrupt
01101	RINT0	McBSP0 receive interrupt
01110	XINT1	McBSP1 transmit interrupt
01111	RINT1	McBSP1 receive interrupt

Source: Courtesy of Texas Instruments.

FP contains eight 32-bit instructions (256 bits) or 32 bytes, each offset address in the table is incremented by $20\,h = 32$.

The reset FP must be at address 0. However, the FPs associated with the other interrupts can be relocated. The relocatable address can be specified by writing this address to the interrupt service table base (ISTB) register of the interrupt service table pointer (ISTP) register, shown in Figure B.7. On reset, ISTB is zero. For relocating the vector table, the ISTP is used; the relocatable address is ISTB plus the offset.

Table 3.6 shows the interrupt selector values needed to choose a specific type of interrupt. The interrupt selector value 01000 is also for EDMA_INT, the enhanced DMA interrupt.

The software defined interrupts INT4–INT15 are associated with a physical interrupt signal using the interrupt multiplex registers IML and IMH. The desired interrupt select values in Table 3.5 are stored in the proper IML or IMH fields for INT4–INT15 [7]. See also the support file C6xdskinterrupt.h.

3.14.2 Selection of XINT0

In most previous examples, the McBSP0 transmit interrupt was chosen. In the communication file C6xdskinit.c, the function Config_Interrupt_Selector is called, which is within the interrupt header support file C6xinterrupts.h. The corresponding interrupt selector number (01100) = 0xC is obtained from C6xinterrupts.h (this 5-bit selector value resides within bits 5 through 9 of the IMH register).

3.14.3 Interrupt Acknowledgment

The signals IACK and INUMx (INUM0 through INUM3) are pins on the C6x that acknowledge an interrupt has occurred and is being processed. The four INUMx signals indicate the number of the interrupt being processed. For example,

```
INUM3 = 1 (MSB), INUM2 = 0, INUM1 = 1, INUM0 = 1 (LSB)
```

corresponds to $(1011)_b$ = 11, indicating that INT11 is being processed.

The IE11 bit is set to 1 to enable INT11. The interrupt flag register (IFR) can be read to verify that bit IF11 is set to 1 (INT11 enabled). Writing a 1 to a bit in the interrupt set register (ISR) causes the corresponding interrupt flag to be set in IFR; whereas a 0 to a bit in the interrupt clear register (ICR) causes the corresponding interrupt to be cleared.

All interrupts remain pending while the CPU has a pending branch instruction. Since a branch instruction has five delay slots, a loop smaller than six cycles is noninterruptible. Any pending interrupt will be processed as long as there are no pending branches to be completed. Additional information can be found in Ref. 6.

3.15 MULTICHANNEL BUFFERED SERIAL PORTS

Two multichannels buffered serial ports (McBSPs) are available. They provide an interface to inexpensive (industry standard) external peripherals. McBSPs have features such as full-duplex communication, independent clocking and framing for receiving and transmitting, and direct interface to AC97 and IIS compliant devices. It allows several data sizes between 8 and 32 bits. Clocking and framing associated with the McBSPs for input and output can be found in Ref. 7.

External data communication can occur while data are being moved internally. Figure 3.4 shows an internal block diagram of a McBSP. The data transmit (DX) and the data receive (DR) pins are used for data communication. Control information (clocking and frame synchronization) is through CLKX, CLKR, FSX, and FSR. The CPU or DMA controller reads data from the data receive register (DRR) and writes data to be transmitted to the data transmit register (DXR). The transmit shift register (XSR) shifts these data to DX. The receive shift register (RSR) copies the data received on DR to the receive buffer register (RBR). The data in RBR are then copied to DRR to be read by the CPU or the DMA controller.

Other registers—serial port control register (SPCR), receive/transmit control register (RCR/XCR), receive/transmit channel enable register (RCER/XCER), pin control register (PCR), and sample rate generator register (SRGR)—support further data communication [7].

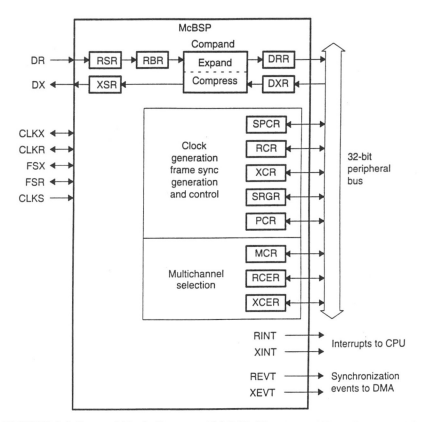

FIGURE 3.4. Internal block diagram of McBSP (Courtesy of Texas Instruments).

3.16 DIRECT MEMORY ACCESS

Direct memory access (DMA) allows for the transfer of data to and from internal memory or external devices without intervention from the CPU [7]. Four DMA channels can be configured independently for data transfer. An additional (auxiliary) channel is available for DMA with the host port interface (HPI). DMA can access on-chip memory and the external memory interface (EMIF). Data of different sizes can be transferred: 8-bit bytes, 16-bit half-words, and 32-bit words.

A number of DMA registers are used to configure the DMA: address (source and destination), index, count reload, DMA global data, and control registers. The source and destination addresses can be from internal program memory, internal data memory, external memory interface, and internal peripheral bus. DMA transfers can be triggered by interrupts from internal peripherals as well as from external pins.

For each resource, each DMA channel can be programmed for priorities with the CPU. Between the four DMA channels, channel 0 has the highest priority and channel 3 has the lowest priority. Each DMA channel can be made to start initiating block transfer of data independently. A block can contain a number of frames. Within each frame can be many elements. Each element is a single data value. The DMA count reload register contains the value to specify the frame count (16 MSBs) and the element count (16 LSBs). An enhanced DMA (EDMA) is also available with 16 independently programmable channels.

3.17 MEMORY CONSIDERATIONS

3.17.1 Data Allocation

Blocks of code and data can be allocated in memory within sections specified in the linker command file. These sections can be either initialized or uninitialized. Initialized or uninitialized sections, except .text, cannot be allocated into internal program memory. The initialized sections are:

1. .cinit: for global and static variables
2. .const: for global and static constant variables
3. .switch: contains jump tables for large switch statements
4. .text: for executable code and constants

The uninitialized sections are:

1. .bss: for global and static variables
2. .far: for global and static variables declared far
3. .stack: allocates memory for the system stack
4. .sysmem: reserves space for dynamic memory allocation used by the malloc, calloc, and realloc functions

The linker can be used to place sections, such as, text in fast internal memory for most efficient operation.

3.17.2 Data Alignment

The C6x always accesses aligned data which allows it to address bytes, half-words, and words (32 bits). The data format consists of four byte boundaries, two half-word boundaries, and one word boundary. For example, to assign a 32-bit load with LDW, the address must be aligned with a word boundary so that the lower 2 bits of the address are zero. Otherwise, incorrect data can be loaded. A double-word (64 bits) also can be accessed. Both .S1 and .S2 can be used to execute the double-word instruction LDDW to load two 64-bit double words, for a total of 128 bits per cycle.

3.17.3 Pragma Directives

The pragma directives tell the compiler to consider certain functions. Pragmas include DATA_ALIGN, DATA_SECTION, and so on. The DATA_ALIGN pragma has the syntax

```
#pragma DATA_ALIGN (symbol,constant);
```

which aligns *symbol* to a boundary. The constant is a power of 2. This pragma directive is used later in conjunction with FFT examples to align data in memory.

The DATA_SECTION pragma has the following syntax:

```
#pragma DATA_SECTION (symbol,"my_section");
```

which allocates space for *symbol* in the section named *my_section*.

Another useful pragma directive,

```
# pragma MUST_ITERATE (20,20)
```

tells the compiler that the loop following will execute 20 times (minimum and maximum of 20 times).

3.17.4 Memory Models

The compiler generates a small memory model code by default. Every data object is handled as if declared *near* unless it is specifically declared *far*. If the DATA_SECTION pragma is used, the object is specified as a *far* variable.

How run-time support functions are called can be controlled by the option –mr0 with the run-time support data and calls *near*, or by the option –mr1 with the run-time support data and calls *far*. Using the *far* method to call functions does not imply that those functions must reside in off-chip memory.

Large-memory models can be generated with the linker options –mlx (x = 0 to 4). If no level is specified, data and functions default to *near*. These models can be used if calling a function that is more than 1 M word away.

3.18 FIXED- AND FLOATING-POINT FORMAT

Some fixed-point considerations are reviewed in Appendix C.

3.18.1 Data Types

Some data types are:

1. *short*: of size 16 bits represented as 2's complement with a range from -2^{15} to $(2^{15} - 1)$

2. *int* or *signed int*: of size 32 bits represented as 2's complement with a range from -2^{31} to $(2^{31} - 1)$

3. *float*: of size 32 bits represented as IEEE 32-bit with a range from $2^{-126} = 1.175494 \times 10^{-38}$ to $2^{+128} = 3.40282346 \times 10^{38}$

4. *double*: of size 64 bits represented as IEEE 64-bit with a range from $2^{-1022} = 2.22507385 \times 10^{-308}$ to $2^{+1024} = 1.79769313 \times 10^{+308}$

Data types such as short for fixed-point multiplication can be more efficient (fewer cycles) than using int. Use of const can also increase code performance.

3.18.2 Floating-Point Format

With a much wider dynamic range in a floating-point processor, scaling is not an issue. A floating-point number can be represented using single precision (SP) with 32 bits or double precision (DP) with 64 bits, as shown in Figure 3.5. In single-precision format, bit 31 represents the sign bit, bits 23 through 30 represent the exponent bits, and bits 0 through 22 represent the fractional bits, as shown in Figure 3.5*a*. Numbers as small as 10^{-38} and as large as 10^{+38} can be represented. In double-precision format, more exponent and fractional bits are available, as shown in Figure 3.5*b*. Since 64 bits are represented, a pair of registers is used. Bits 0 through 31 of the first register pair represent the fractional bits. Bits 0 through 19 of the second register pair also represent the fractional bits, with bits 20 through 30 representing the exponent bits, and bit 31 the sign bit. As a result, numbers as small as 10^{-308} and as large as 10^{+308} can be represented.

Instructions ending in either SP or DP represent single and double precision, respectively. Some of the floating-point instructions (available on the C67x floating-point processor) have more latencies than do fixed-point instructions. For example, the fixed-point multiplication *MPY* requires one delay or *NOP*, whereas the single-precision *MPYSP* requires three delays, and the double-precision instruction *MPYDP* requires nine delays.

The single-precision floating-point instructions ADDSP and MPYSP have three delay slots and take four cycles to complete execution. The double-precision instruc-

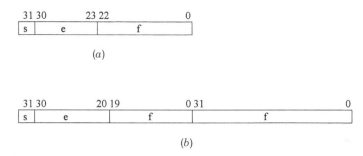

FIGURE 3.5. Data format: (*a*) single precision; (*b*) double precision.

tions ADDDP and MPYDP have six and nine delay slots, respectively. However, the floating-point double-word load instruction LDDW (with four delay slots as with the fixed-point LDW) can load 64 bits. Two LDDW instructions can execute in parallel through both units .S1 and .S2 to load a total of 128 bits per cycle.

A single-precision floating-point value can be loaded into a single register, whereas a double-precision floating-point value is a 64-bit value that can be loaded into a register pair such as A1:A0, A3:A2, . . . , B1:B0, B3:B2, . . . The least significant 32 bits are loaded into the even register pair, and the most significant 32 bits are loaded into the odd register pair.

One may need to weigh the pros and cons of dynamic range and accuracy with possible degradation in speed when using floating-point types of instructions.

3.18.3 Division

The floating-point C6711 processor has a single-precision reciprocal instruction RCPSP. A division operation can be performed by taking the reciprocal of the denominator and multiplying the result by the numerator [6]. There are no fixed-point instructions for division. Code is available to perform a division operation by using the fixed-point processor to implement a Newton–Raphson equation.

3.19 CODE IMPROVEMENT

Several code optimization schemes are discussed in Chapter 8 using both fixed- and floating-point implementations and ASM code.

3.19.1 Intrinsics

C code can be optimized further by using many of the intrinsics available from the run-time library support file. Intrinsic functions are similar to run-time support library functions. Intrinsics are available to multiply, to add, to find the reciprocal of a square root, and so on. For example, in lieu of using the asterisk operator to multiply, the intrinsic _mpy can be used. Intrinsics are special functions that map directly to inline C6x instructions. For example,

```
int _mpy()
```

is equivalent to the assembly instruction MPY, to multiply the 16 LSBs of two numbers. The intrinsic function

```
int _mpyh()
```

is equivalent to the assembly instruction MPYH to multiply the 16 MSBs of two numbers.

3.19.2 Trip Directive for Loop Count

The linear assembly directive .$trip$ is used to specify the number of times a loop iterates. If the exact number is known and used, the linear assembler optimizer can produce pipelined code (discussed in Chapter 8) and redundant loops are not generated. This can improve both code size and execution time. A .$trip$ count specification, even if it is not the exact value, may improve performance: for example, when the actual number of iterations is a multiple of the specified value. The intrinsic function _$nassert$() can be used in a C program in lieu of .$trip$. Example 3.1 illustrates the use of _$nassert$() in the dot product example.

3.19.3 Cross-Paths

Data and address cross-path instructions are used to increase code efficiency. The instruction

```
MPY   .M1x   A2,B2,A4
```

illustrates a data cross-path that multiplies the two sources A2 and B2 from two different sides, A and B, with the result in A4. If the result is in the B register file, a 2x cross-path is used with the instruction

```
MPY   .M2x   A2,B2,B4
```

with the result in B4. The instruction

```
LDW   .D1T2   *A2,B2
```

illustrates an address cross-path. It loads the content in register A2 (from a register file A) into register B2 (register file B). Only two cross-paths are available on the C6x, so no more than two instructions using cross-paths are allowed within a cycle.

3.19.4 Software Pipelining

Software pipelining uses available resources to obtain efficient pipelining code. The aim is to use all eight functional units within one cycle. However, substantial coding effort is required using the software pipelining technique. There are three stages to a pipelined code:

1. Prolog
2. Loop kernel (or loop cycle)
3. Epilog

The first stage, prolog, contains instructions to build the second-stage loop cycle, and the epilog stage (last stage) contains instructions to finish all loop iterations. Software pipelining is used by the compiler when optimization option level –o2 or –o3 is invoked. The most efficient software pipelined code has loop trip counters that count down: for example,

```
for (i = N; i != 0; i--)
```

A dot product example with word-wide hand-coded pipelined code results in ($N/2$) + 8 cycles to obtain the sum of two arrays, with N numbers in each array. This translates to 108 cycles to find the sum of products of 200 numbers, as illustrated in Chapter 8. This efficiency is obtained using instructions such as LDW to load a 32-bit word, and multiplying the lower and higher 16-bit numbers separately with the two instructions *mpy* and *mpyh*, respectively.

Removing the epilog section can also reduce the code size. The available options –msn (n = 0,1,2) directs the compiler to favor code size reduction over performance. Producing a hand-coded software pipelined code can be obtained by first drawing a dependency graph and setting up a scheduling table [8]. In Chapter 8 we discuss software pipelining in conjunction with code efficiency.

3.20 CONSTRAINTS

3.20.1 Memory Constraints

Internal memory is arranged through various banks of memory so that loads and stores can occur simultaneously. Since each bank of memory is single-ported, only one access to each bank is performed per cycle. Two memory accesses per cycle can be performed if they do not access the same bank of memory. If multiple accesses are performed to the same bank of memory (within the same space), the pipeline will stall. This causes additional cycles for execution to complete.

3.20.2 Cross-Paths Constraints

Since there is one cross-path in each side of the two data paths, there can be at most two instructions per cycle using cross-paths. The following code segment is valid since both available cross-paths are utilized:

```
   ADD  .L1x  A1,B1,A0
|| MPY  .M2x  A2,B2,B3
```

whereas the following is not valid since one cross-path is used for both instructions:

```
   ADD   .L1x  A1,B1,A0
|| MPY   .M1x  A2,B2,A3
```

The x associated with the functional unit designates a cross-path.

3.20.3 Load/Store Constraints

The address register to be used must be on the same side as the .D unit. The following code segment is valid:

```
   LDW   .D1  *A1,A2
|| LDW   .D2  *B1,B2
```

whereas the following is not valid:

```
   LDW   .D1  *A1,A2
|| LDW   .D2  *A3,B2
```

Furthermore, loading and storing cannot be from the same register file. A load (or store) using one register file in parallel with another load (or store) must use a different register file. For example, the following code segment is valid:

```
   LDW   .D1  *A0,B1
|| STW   .D2  A1,*B2
```

The following is also valid:

```
   LDW   .D1  *A0,B1
|| LDW   .D2  *B2,A1
```

However, the following is not valid:

```
   LDW   .D1  *A0,A1
|| STW   .D2  A2,*B2
```

3.20.4 Pipelining Effects with More Than One EP within an FP

Table 3.3 shows a previous pipeline operation representing eight instructions in parallel within one fetch packet (FP). Table 3.7 shows the pipeline operation when there are more than one execute packet (EP) within an FP.

Consider the operation of six fetch packets (FP1 through FP6) through the pipeline. FP1 contains three execute packets, and FP2, FP3, ..., FP6 each contains

TABLE 3.7   Pipelining with Stalling Effects

					Clock Cycle						
1	2	3	4	5	6	7	8	9	10	11	12
PG	PS	PW	PR	DP	DC	E1	E2	E3	E4	E5	E6
					DP	DC	E1	E2	E3	E4	E5
						DP	DC	E1	E2	E3	E4
	PG	PS	PW	PR	X	X	DP	DC	E1	E2	E3
	PG	PS	PW	X	X	PR	DP	DC	E1	E2
		PG	PS	X	X	PW	PR	DP	DC	E1
			PG	X	X	PS	PW	PR	DP	DC
				X	X	PG	PS	PW	PR	DP

one execute packet. In cycles 2 through 5, FP2 through FP5, each starts its program fetch phase. When the CPU detects that FP1 contains more than one EP, it forces the pipeline to stall so that EP2 and EP3, within FP1, can each start its dispatching phase in cycles 6 and 7, respectively. Each instruction within an FP has a "p" bit to specify whether that instruction is in parallel with a subsequent instruction (if a 1, as shown in Figure 3.3).

During clock cycles 1 through 4, a program fetch phase occurs. The three EPs within the same fetch packet cause a stall in the pipeline. This allows the DP phase to start at cycle 6 (not at cycle 5) for EP2 and at cycle 7 for EP3. The subsequent fetch packet (FP2) with only one EP (with all eight instructions in parallel) is stalled so that each of the three EPs in the previous FP (FP1) can go through the DP phase. As a result, while the fetch phase for FP2 starts at cycle 2, its DP phase does not start until cycle 8. The third fetch packet (FP3), also with only one EP, starts its fetch stage at cycle 3, but its DP phase does not start until cycle 9, due to the pipeline stall.

The pipeline then stalls in cycles 6 and 7, as indicated with an "X". Once EP3 (within FP1) continues onto its decoding phase in cycle 8, the pipeline is released. FP2 can now continue to its dispatching phase in cycle 8. Since FP3 through FP6 also were stalled, each can now resume its program fetch phase in cycle 8.

Hence, with the three EPs within one FP, the pipeline stalls for two cycles. Table 3.7 illustrates the stalling pipeline effects. A pipeline stall would also take place in the event that the first FP had four EPs, each with two parallel instructions.

3.21 TMS320C64x PROCESSOR

Another member of the C6000 family of processors is the C64x, which can operate at a much higher clock rate, reaching the gigahertz range. Operating at 750 MHz with eight instructions per cycle, this translates to 6000 million instructions per second (MIPS).

The C64x is based on the architecture VELOCITI.2, which is an extension of VELOCITI [8]. Some of its features include a larger memory and twice as many registers, for a total of sixty-four 32-bit registers. The extra registers allow for packed data types to support four 8-bit or two 16-bit operations associated with one 32-bit register; hence increasing parallelism. For example, the instruction MPYU4 performs four 8-bit multiplications within a single instruction cycle time. Several special-purpose instructions have also been added to handle many operations encountered in wireless and digital imaging applications, where 8-bit data processing is common. In addition, the .M unit (for multiply operations) can also handle shift and rotate operations. Similarly, the .D unit (for data manipulation) can also handle logical operations.

The C64x is a fixed-point processor. Existing instructions are available to more units. Double-word load (LDDW) and store (STDW) instructions can access 64 bits of data, with up to two double-word load or store instructions per cycle (read or write 128 bits per cycle).

A few instructions have been added for the C64x processor. For example, the instruction

```
BDEC    LOOP,B0
```

decrements a counter B0 and performs a conditional (based on B0) branch to LOOP. The branch decision is *before* the decrement; with the branch decision based on a negative number (not on whether the number is zero). This multitask instruction resembles the syntax used in the C3x and C4x family of processors.

Furthermore, with the intrinsic C function _dotp2, it can perform two 16×16 multiplies and adds the products together to further reduce the number of cycles. This intrinsic function in C has the corresponding assembly function DOTP2. With two multiplier units, four 16×16 multiplies per cycle can be performed, double the rate of the C62x or C67x. At 750 MHz, this corresponds to 3 billion multiply operations per second; or 6 billion 8×8 multiplies per second.

3.22 PROGRAMMING EXAMPLES USING C, ASSEMBLY, AND LINEAR ASSEMBLY

Six programming examples are discussed in this section. The first example illustrates use of the intrinsic function _nassert to increase the efficiency of the dot product example. The other five examples illustrate both assembly code and linear assembly code implementation: a C program calling an assembly function, a C program calling a linear assembly function, and an assembly-coded program calling an assembly-coded function. The focus here is on illustrating the syntax of both assembly and linear assembly code, not necessarily to produce optimized code. We discuss further optimization techniques in Chapter 8 in conjunction with code efficiency and software pipelining.

Example 3.1: Efficient Dot Product (dotpopt)

This example uses the intrinsic function _nassert in the dot product example introduced in Chapter 1. Figure 3.6 shows a listing of the program dotpopt.c, which calls the C function dotpfunc.c listed in Figure 3.7. This function produces more efficient code, with _nassert used for the alignment of the incoming pointers as constant pointers. This provides additional information to the compiler about the loop.

Verify that using compiler options −g and −o3, the number of cycles associated with profiling the function dotpfunc.c is reduced from 100 (without the intrin-

```
//dotpopt.c Optimized dot product of two arrays

#include    <stdio.h>
#include    "dotp4.h"
#define     count 4

short x[count] = {x_array};                 //declare 1st array
short y[count] = {y_array};                 //declare 2nd array
volatile int result = 0;                    //result

main()
{
  result = dotpfunc(x,y,count);             //call optimized function
  printf("result = %d decimal \n", result); //print result
}
```

FIGURE 3.6. Dot product program calling function with _nassert intrinsic (dotpopt.c).

```
//dotpfunc.c Optimized dot product function

int dotpfunc(const short *a, const short *b, int ncount)
{
    int sum = 0;
    int i;

  _nassert((int)(a)%4 == 0);
  _nassert((int)(b)%4 == 0);
  _nassert((int)(ncount)%4 == 0);

  for ( i = 0; i < ncount; i++)
  {
      sum += (a[i] * b[i]);          //sum of products
  }
  return (sum);                      //return sum as result
}
```

FIGURE 3.7. C-called function for a dot product using _nassert (dotpfunc.c).

sics functions) to 71 (with intrinsics). Using the options $-g$, $-pm$, and $-o3$, the number of cycles is further reduced to 30. The $-pm$ option uses program level optimization, with the source files compiled into one intermediate file. The results with this option can be compared to the results obtained with the function dotp in Example 1.3.

In Chapter 8 we use optimization techniques associated with the dot product example, using two arrays each with N numbers. We show that the number of cycles can be reduced to $7 + (N/2) + 1$ with a fixed-point implementation, or 108 cycles using 200 numbers in each array. For a floating-point implementation, we obtain 124 cycles (see Table 8.4).

Example 3.2: Sum of n + (n − 1) + (n − 2) + . . . + 1 Using C Calling Assembly Function (sum)

This example illustrates a C program calling an assembly function. The C source program *sum.c* (Figure 3.8) calls the assembly-coded function *sumfunc.asm*

```
//Sum.c Finds n+(n-1)+...+1. Calls assembly function sumfunc.asm

#include <stdio.h>

main()
{
 short n=6;                 //set value
 short result;             //result from asm function

 result = sumfunc(n);      //call assembly function sumfunc
 printf("sum = %d", result); //print result from asm function
}
```

FIGURE 3.8. C program that calls an ASM function to find $n + (n − 1) + (n − 2) + . . . + 1$ (sum.c).

```
;Sumfunc.asm Assembly function to find n+(n-1)+...+1

          .def      _sumfunc ;function called from C
_sumfunc: MV    .L1  A4,A1    ;setup n as loop counter
          SUB   .S1  A1,1,A1  ;decrement n

LOOP:     ADD   .L1  A4,A1,A4 ;accumulate in A4
          SUB   .S1  A1,1,A1  ;decrement loop counter
    [A1]  B     .S2  LOOP     ;branch to LOOP if A1#0
          NOP        5        ;five NOPs for delay slots
          B     .S2  B3       ;return to calling routine
          NOP        5        ;five NOPs for delay slots
          .end
```

FIGURE 3.9. ASM function called from C in the project sum (sumfunc.asm).

(Figure 3.9). It implements the sum of $n + (n - 1) + (n - 2) + \ldots + 1$. The value of n is set in the main C program. It is passed through register A4 (by convention). For example, the address of more than one value can be passed to the assembly function through A4, B4, A6, . . . The resulting sum from the assembly function is returned to `result` in the C program, which then prints this resulting sum.

The assembly function's name is preceded by an underscore (by convention). The value n in register A4 in the `asm` function is moved to register A1 to set A1 as a loop counter. A1 is then decremented. A loop section of code starts with the label or address LOOP and ends with the first branch statement B. The first addition adds $n + (n - 1)$ with the result in A4. A1 is again decremented to $(n - 2)$. The branch statement is conditional based on register A1 (only A1, A2, B0, B1, and B2 can be used as conditional registers), and since A1 is not zero, branching takes place and execution returns to the instruction at the address LOOP, where $A4 = n + (n - 1)$ is added to $A1 = (n - 2)$. This process continues until register $A1 = 0$.

The second branch instruction is to the returning address B3 (by convention) of the C calling program. The resulting sum is contained or accumulated in A4, which is passed to `result` in the C program. The five NOPs (no operation) are to account for the five delay slots associated with a branch instruction.

The functional units .S and .L selected are shown but are not required in the program. They can be useful for debugging and analyzing which of the functional units are used in order to improve on the efficiency of the program. Similarly, the two colons after the label LOOP and the function name are not required.

Build and run this project as **sum**. With a value of n set to 6 in the C program, verify that sum and its value of 21 are printed.

Example 3.3: Factorial of a Number Using C Program Calling Assembly Function (`factorial`)

This example finds the factorial of a number $n \leq 7$ with $n! = n(n - 1)(n - 2) \ldots (1)$. It further illustrates the syntax of assembly code. It is very similar to Example 3.2. The value of n is set in the C source program `factorial.c`, shown in Figure 3.10, which calls the assembly function `factfunc.asm`, shown in Figure 3.11. It is instructive to read the comments.

Register A1 is again set as a loop counter. Within the loop section of code starting with at the address LOOP, the first multiply is $n(n - 1)$ and accumulates in register A4. The initial value of n is passed to the `asm` function through A4. The MPY instruction has one delay slot, hence the NOP following it. Processing continues within the loop section of code until $A1 = 0$. Note that the functional units are not specified in this program. The resulting factorial is returned to the calling C program through A4.

Build and run this project as **factorial**. Verify that $factorial$ and its value of 5040 (7!) are printed. Note that the maximum value of n is 7, since 8! is greater than 2^{15}.

```
//Factorial.c Finds factorial of n. Calls function factfunc.asm

#include <stdio.h>                //for print statement

void main()
{
 short n=7;                       //set value
 short result;                    //result from asm function

 result = factfunc(n);           //call assembly function factfunc
 printf("factorial = %d", result); //print result from asm function
}
```

FIGURE 3.10. C program that calls an ASM function to find the factorial of a number (factorial.c).

```
;Factfunc.asm Assembly function called from C to find factorial

          .def   _factfunc          ;asm function called from C
_factfunc: MV    A4,A1              ;setup loop count in A1
           SUB   A1,1,A1            ;decrement loop count
LOOP:      MPY   A4,A1,A4           ;accumulate in A4
           NOP                      ;for 1 delay slot with MPY
           SUB   A1,1,A1            ;decrement for next multiply
     [A1]  B     LOOP               ;branch to LOOP if A1 # 0
           NOP   5                  ;five NOPs for delay slots
           B     B3                 ;return to calling routine
           NOP   5                  ;five NOPs for delay slots
          .end
```

FIGURE 3.11. ASM function called from C that finds the factorial of a number (factfunc.asm).

Example 3.4: Dot Product Using Assembly Program Calling Assembly Function (dotp4a)

This example takes the sum of products of two arrays, each array with four numbers. See also Example 1.3, which implements it using only C code, and Examples 3.2 and 3.3, which introduced the syntax of assembly code. Figure 3.12 shows a listing of the assembly program dotp4a_init.asm, which initializes the two arrays of numbers and calls the assembly function dotp4afunc.asm (Figure 3.13) which takes the sum of products of the two arrays. It also sets a return address through register B3 and the result address to A0. The addresses of the two arrays and the size of the array are passed to the function dotp4afunc.asm through registers A4, A6, and B4, respectively. The result from the called function is "sent back" through A4. The resulting sum of product is stored in memory whose address is

```
;Dotp4a_init.asm ASM program to init variables. Calls dotp4afunc.asm

              .def          init          ;starting address
              .ref          dotp4afunc    ;called ASM function
              .text                       ;section for code
x_addr        .short        1,2,3,4       ;numbers in x array
y_addr        .short        0,2,4,6       ;numbers in y array
result_addr .short         0             ;initialize sum of products

init          MVK    result_addr,A4       ;A4 = lower 16-bit addr -->A4
              MVKH   result_addr,A4       ;A4 = higher 16-bit addr-->A4
              MVK    0,A3                 ;A3 = 0
              STH    A3,*A4               ;init result to 0
              MVK    x_addr,A4            ;A4 = 16 MSBs address of x
              MVK    y_addr,B4            ;B4 = 16 LSBs address of y
              MVKH   y_addr,B4            ;B4 = 16 MSBs address of y
              MVK    4,A6                 ;A6 = size of array
              B      dotp4afunc           ;branch to function dotp4afunc
              MVK    ret_addr,b3          ;B3 = return addr from dotp4a
              MVKH   ret_addr,b3          ;B3 = return addr from dotp4a
              NOP    3                    ;3 more delay slots(branch)

ret_addr      MVK    result_addr,A0       ;A0 = 16 LSBs result_addr
              MVKH   result_addr,A0       ;A0 = 16 MSBs result_addr
              STW    A4,*A0               ;store result
wait          B      wait                 ;wait here
              NOP    5                    ;delay slots for branch
```

FIGURE 3.12. ASM program calling ASM function to find the sum of products (dotp4a_init.asm).

result_addr. The instruction STW stores the resulting sum of products (in A4) in memory pointed by A0. Register A0 serves as a pointer with the address result_addr.

The starting address of the calling ASM program is defined as init. The vector file *vectors_dotp4a.asm* (Figure 3.14) specifies a branch to that entry address. The called ASM function *dotp4afunc.asm* calculates the sum of products. The loop count value was moved to A1 since A6 cannot be used as a conditional register (only A1, A2, B0, B1, B2 can be used). The two LDH instructions load (half-word of 16 bits) the addresses of the two arrays starting at *x_addr* and *y_addr* into registers A2 and B2, respectively. For example, the instruction

```
LDH  *B4++,B2
```

loads the content in memory (the first value in the second array starting at *y_address*) pointed by B4 (the address of the second array) into B2. Then

```
;Dotp4afunc.asm Multiply two arrays. Called from dotp4a_init.asm
;A4=x address,B4=y address,A6=count(size of array),B3=return address

            .def        dotp4afunc  ;dot product function
            .text                   ;text section
dotp4afunc  MV          A6,A1       ;move loop count -->A1
            ZERO        A7          ;init A7 for accumulation

loop        LDH         *A4++,A2    ;A2=(x. A4 as address pointer
            LDH         *B4++,B2    ;B2=(y). B4 as address pointer
            NOP         4           ;4 delay slots for LDH
            MPY         A2,B2,A3    ;A3 = x * y
            NOP                     ;1 delay slot for MPY
            ADD         A3,A7,A7    ;sum of products in A7
            SUB         A1,1,A1     ;decrement loop counter
     [A1]   B           loop        ;branch back to loop till A1=0
            NOP         5           ;5 delay slots for branch

            MV          A7,A4       ;A4=result A4=return register
            B           B3          ;return from func to addr in B3
            NOP         5           ;5 delay slots for branch
```

FIGURE 3.13. ASM function called from an ASM program to find the sum of products (`dotp4afunc.asm`).

```
;vectors_dotp4a.asm Vector file for dotp4a project

            .ref        init        ;starting addr in init file
            .sect       "vectors"   ;in section vectors
rst:        mvkl .s2    init,b0     ;init addr 16 LSB --->B0
            mvkh .s2    init,b0     ;init addr 16 MSB --->B0
            b           b0          ;branch to addr init
            nop
            nop
            nop
            nop
            nop
```

FIGURE 3.14. Vector file that specifies the entry address in the calling ASM program for the sum of products (`vectors_dotp4a.asm`).

register B4, used as a pointer, is postincremented to the next-higher address in memory that contains the second value in the second array. Register A7 is used to accumulate and move the sum of products to register A4, since the result is passed to the calling function through A4.

Support files for this project include (no library file is necessary):

1. dotp4a_init.asm
2. dotp4afunc.asm
3. vectors_dotp4a.asm

Build and run this project as **dotp4a**. Modify the Linker Option (Project → Options) to select "No Autoinitialization." Otherwise, the warning "entry point symbol _c_int00 undefined" is displayed when this project is built (it can be ignored). This is because the "conventional" entry point is not used in this project with no main function in C.

Set a breakpoint at the first branch instruction in the program dotp4a_init.asm:

```
B dotp4afunc
```

Select View → Memory and set address to result_addr and use 16-bit signed integer. Right-click on the memory window and deselect "Float in Main Window." This allows you to have a better display of the Memory window while viewing the source file dotp4a_init.asm.

Select Run. Execution stops at the set breakpoint. The content in memory at the address result_addr is zero (the called function dotp4afunc.asm is not yet executed). Run again, then halt (since execution is within the infinite wait loop instruction):

```
wait  B  wait  ;wait here
```

Verify that the resulting sum of products is now 40. Note that A0 contains the result address (result_addr). View → CPU Registers → Core Registers and verify this address (in hex). Figure 3.15 shows a CCS display of this project. Note from the disassembly file that execution was halted at the infinite wait loop.

Example 3.5: Dot Product Using C Function Calling Linear Assembly Function (dotp4clasm)

Figure 3.16 shows a listing of the C program dotp4clasm.c, which calls the linear assembly function dotp4clasmfunc.sa (Figure 3.17). Example 1.3 introduced the dot product implementation using C code only. The previous three examples introduced the syntax of assembly-coded programs.

The section of code invoked by the linear assembler optimizer starts and ends with the linear assembler directives .cproc and .endproc, respectively. The name of the linear assembly function called is preceded by an underscore since the calling function is in C. The directive .ref (or .def) references (defines) the function.

Functional units are optional as in an assembly-coded program. Registers a, b, prod and sum are defined by the linear assembler directive .reg. The addresses

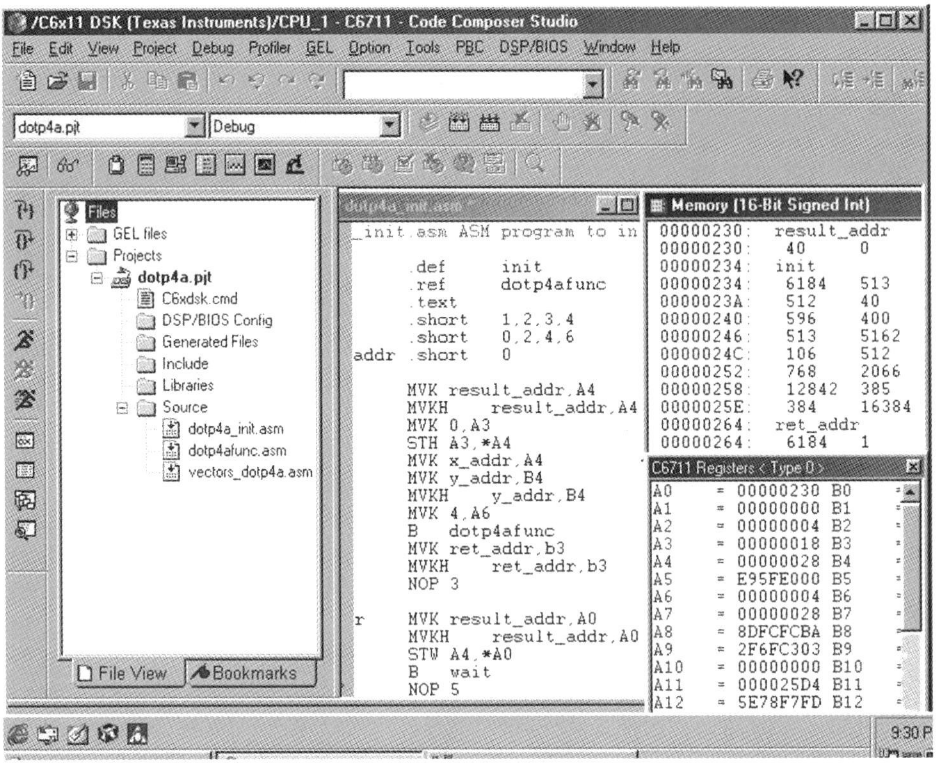

FIGURE 3.15. CCS windows for the sum of products in the project `dotp4a`.

```
//Dotp4clasm.c Multiplies two arrays using C calling linear ASM func

short dotp4clasmfunc(short *a,short *b,short ncount);   //prototype
#include <stdio.h>                         //for printing statement
#include "dotp4.h"                         //arrays of data values
#define   count 4                          //number of data values
short x[count] = {x_array};               //declare 1st array
short y[count] = {y_array};               //declare 2nd array
volatile int result = 0;                   //result

main()
{
  result = dotp4clasmfunc(x,y,count);      //call linear ASM func
  printf("result = %d decimal \n", result); //print result
}
```

FIGURE 3.16. C program calling a linear ASM function to find the sum of products (`dotp4clasm.c`).

```
;Dotp4clasmfunc.sa Linear assembly function to multiply two arrays
                  .ref     _dotp4clasmfunc  ;ASM func called from C
_dotp4clasmfunc:  .cproc   ap,bp,count      ;start section linear asm
                  .reg     a,b,prod,sum     ;asm optimizer directive

                  zero     sum              ;init sum of products
loop:             ldh      *ap++,a          ;pointer to 1st array->a
                  ldh      *bp++,b          ;pointer to 2nd array->b
                  mpy      a,b,prod         ;product= a*b
                  add      prod,sum,sum     ;sum of products-->sum
                  sub      count,1,count    ;decrement counter
   [count]        b        loop             ;loop back if count # 0

                  .return  sum              ;return sum as result
                  .endproc                  ;end linear asm function
```

FIGURE 3.17. Linear ASM function called from C to find the sum of products (dotp4clasmfunc.sa).

of the two arrays x and y and the size of the array (count) are passed to the linear assembly function through the registers *ap*, *bp*, and *count*. Both *ap* and *bp* are registers used as pointers, as in C code. The instruction field is seen to be as in an assembly-coded program and the subsequent field uses a syntax as in C programming. For example, the instruction

```
loop:    ldh     *ap++,a
```

(the first time through the loop section of code) loads the content in memory, whose address is specified by register ap, into register a. Then the pointer register ap is postincremented to point to the next-higher memory address, pointing at the memory location containing the second value of x within the x array. The value of the sum of products is accumulated in *sum*, which is returned to the C calling program.

Build and run this project as **dotp4clasm**. Verify that the following is printed: result = 40. You may wish to profile the linear assembly code function and compare its execution time with the C-coded version in Example 1.3.

Example 3.6: Factorial Using C Calling a Linear Assembly Function (`factclasm`)

Figure 3.18 shows a listing of the C program `factclasm.c`, which calls the linear ASM function `factclasmfunc.sa` (Figure 3.19) to calculate the factorial of a number less than 8. See also Example 3.3, which finds the factorial of a number using a C program that calls an ASM function. Example 3.5 illustrates a C

```
//Factclasm.c Factorial of number. Calls linear ASM function

#include <stdio.h>                        //for print statement

void main()
{
 short number = 7;                         //set value
 short result;                             //result of factorial

 result = factclasmfunc(number);     //call ASM function factlasmfunc
 printf("factorial = %d", result); //print from linear ASM function
}
```

FIGURE 3.18. C program that calls a linear ASM function to find the factorial of a number (`factclasm.c`).

```
;Factclasmfunc.sa Linear ASM function called from C to find factorial

                .ref    _tactclasmfunc  ;Linear ASM func called from C
_factclasmfunc: .cproc  number          ;start of linear ASM function
                .reg    a,b             ;asm optimizer directive
                mv      number,b        ;set-up loop count in b
                mv      number,a        ;move number to a
                sub     b,1,b           ;decrement loop counter

loop:           mpy     a,b,a           ;n(n-1)
                sub     b,1,b           ;decrement loop counter
  [b]           b       loop            ;loop back to loop if count # 0
                .return a               ;result to calling function
                .endproc                ;end of linear asm function
```

FIGURE 3.19. Linear ASM function called from C that finds the factorial of a number (`factclasmfunc.sa`).

program calling a linear ASM function to find the sum of products and is instructive for this project. Examples 3.3 and 3.5 cover the essential background for this project.

Support files for this project include `factclasm.c`, `factclasmfunc.sa`, `vectors`, `rts6701.lib`, and `C6xdsk.cmd`. Build and run this project as **factclasm**. Verify that the result of 7! is printed, or `factorial = 5040`.

REFERENCES

1. R. Chassaing and D. W. Horning, *Digital Signal Processing with the TMS320C25*, Wiley, New York, 1990.

2. R. Chassaing, *Digital Signal Processing Laboratory Experiments Using C and the TMS320C31 DSK*, Wiley, New York, 1999.

3. R. Chassaing, *Digital Signal Processing with C and the TMS320C30*, Wiley, New York, 1992.

4. R. Chassaing and P. Martin, Parallel processing with the TMS320C40, *Proceedings of the 1995 ASEE Annual Conference*, June 1995.

5. R. Chassaing and R. Ayers, Digital signal processing with the SHARC, *Proceedings of the 1996 ASEE Annual Conference*, June 1996.

6. *TMS320C6000 CPU and Instruction Set*, SPRU189F, Texas Instruments, Dallas, TX, 2000.

7. *TMS320C6000 Peripherals*, SPRU190D, Texas Instruments, Dallas, TX, 2001.

8. *TMS320C6000 Programmer's Guide*, SPRU198D, Texas Instruments, Dallas, TX, 2000.

9. *TMS320C6000 Assembly Language Tools User's Guide*, SPRU186G, Texas Instruments, Dallas, TX, 2000.

10. *TMS320C6000 Optimizing Compiler User's Guide*, SPRU187G, Texas Instruments, Dallas, TX, 2000.

11. *TMS320C6211 Fixed-Point Digital Signal Processor—TMS320C6711 Floating-Point Digital Signal Processor*, SPRS073C, Texas Instruments, Dallas, TX, 2000.

4

Finite Impulse Response Filters

- Introduction to the z-transform
- Design and implementation of finite impulse response (FIR) filters
- Programming examples using C and TMS320C6x code

The z-transform is introduced in conjunction with discrete-time signals. Mapping from the s-plane, associated with the Laplace transform, to the z-plane, associated with the z-transform, is illustrated. FIR filters are designed with the Fourier series method and implemented by programming a discrete convolution equation. Effects of window functions on the characteristics of FIR filters are covered.

4.1 INTRODUCTION TO THE *Z*-TRANSFORM

The z-transform is utilized for the analysis of discrete-time signals, similar to the Laplace transform for continuous-time signals. We can use the Laplace transform to solve a differential equation that represents an analog filter, or the z-transform to solve a difference equation that represents a digital filter. Consider an analog signal $x(t)$ ideally sampled

$$x_s(t) = \sum_{k=0}^{\infty} x(t)\delta(t - kT) \tag{4.1}$$

where $\delta(t - kT)$ is the impulse (delta) function delayed by kT and $T = 1/F_s$ is the sampling period. The function $x_s(t)$ is zero everywhere except at $t = kT$. The Laplace transform of $x_s(t)$ is

102

$$X_s(s) = \int_0^\infty x_s(t)e^{-st}dt$$

$$= \int_0^\infty \{x(t)\delta(t) + x(t)\delta(t-T) + \cdots\}e^{-st}dt \tag{4.2}$$

From the property of the impulse function

$$\int_0^\infty f(t)\delta(t-kT)dt = f(kT)$$

$X_s(s)$ in (4.2) becomes

$$X_s(s) = x(0) + x(T)e^{-sT} + x(2T)e^{-2sT} + \cdots = \sum_{n=0}^\infty x(nT)e^{-nsT} \tag{4.3}$$

Let $z = e^{sT}$ in (4.3), which becomes

$$X(z) = \sum_{n=0}^\infty x(nT)z^{-n} \tag{4.4}$$

Let the sampling period T be implied; then $x(nT)$ can be written as $x(n)$, and (4.4) becomes

$$X(z) = \sum_{n=0}^\infty x(n)z^{-n} = ZT\{x(n)\} \tag{4.5}$$

which represents the z-transform (ZT) of $x(n)$. There is a one-to-one correspondence between $x(n)$ and $X(z)$, making the z-transform a unique transformation.

Exercise 4.1: ZT of Exponential Function $x(n) = e^{nk}$

The ZT of $x(n) = e^{nk}$, $n \geq 0$ and k a constant, is

$$X(z) = \sum_{n=0}^\infty e^{nk}z^{-n} = \sum_{n=0}^\infty (e^k z^{-1})^n \tag{4.6}$$

Using the geometric series, obtained from a Taylor series approximation

$$\sum_{n=0}^\infty u^n = \frac{1}{1-u} \qquad |u| < 1$$

(4.6) becomes

$$X(z) = \frac{1}{1 - e^k z^{-1}} = \frac{z}{z - e^k} \tag{4.7}$$

for $\left|e^k z^{-1}\right| < 1$ or $|z| > \left|e^k\right|$. If $k = 0$, the ZT of $x(n) = 1$ is $X(z) = z/(z-1)$.

Exercise 4.2: ZT of Sinusoid x(n) = sin nωT

A sinusoidal function can be written in terms of complex exponentials. From Euler's formula $e^{ju} = \cos u + j \sin u$,

$$\sin n\omega T = \frac{e^{jn\omega T} - e^{-jn\omega T}}{2j}$$

Then

$$X(z) = \frac{1}{2j} \sum_{n=0}^{\infty} \left(e^{jn\omega T} z^{-n} - e^{-jn\omega T} z^{-n}\right) \tag{4.8}$$

Using the geometric series as in Exercise 4.1, one can solve for $X(z)$; or the results in (4.7) can be used with $k = j\omega T$ in the first summation of (4.8) and $k = -j\omega T$ in the second, to yield

$$\begin{aligned}
X(z) &= \frac{1}{2j}\left(\frac{z}{z - e^{j\omega T}} - \frac{z}{z - e^{-j\omega T}}\right) \\
&= \frac{1}{2j} \frac{z^2 - ze^{-j\omega T} - z^2 + ze^{j\omega T}}{z^2 - z(e^{-j\omega T} + e^{j\omega T}) + 1} \\
&= \frac{z \sin \omega T}{z^2 - 2z \cos \omega T + 1} \\
&= \frac{Cz}{z^2 - Az - B} \qquad |z| > 1
\end{aligned} \tag{4.9}$$
$$\tag{4.10}$$

where $A = 2\cos \omega T$, $B = -1$, and $C = \sin \omega T$. In Chapter 5 we generate a sinusoid based on this result. We can readily generate sinusoidal waveforms of different frequencies by changing the value of ω in (4.9).

Similarly, using Euler's formula for $\cos n\omega T$ as a sum of two complex exponentials, one can find the ZT of $x(n) = \cos n\omega T = (e^{jn\omega T} + e^{-jn\omega T})/2$, as

$$X(z) = \frac{z^2 - z \cos \omega T}{z^2 - 2z \cos \omega T + 1} \qquad |z| > 1 \tag{4.11}$$

4.1.1 Mapping from *s*-Plane to *z*-Plane

The Laplace transform can be used to determine the stability of a system. If the poles of a system are on the left side of the $j\omega$ axis on the s-plane, a time-decaying system response will result, yielding a stable system. If the poles are on the right side of the $j\omega$ axis, the response will grow in time, making such a system unstable. Poles located on the $j\omega$ axis, or purely imaginary poles, will yield a sinusoidal response. The sinusoidal frequency is represented by the $j\omega$ axis, and $\omega = 0$ represents dc (direct current).

In a similar fashion, we can determine the stability of a system based on the location of its poles on the z-plane associated with the z-transform, since we can find corresponding regions between the s-plane and the z-plane. Since $z = e^{sT}$ and $s = \sigma + j\omega$,

$$z = e^{\sigma T} e^{j\omega T} \tag{4.12}$$

Hence, the magnitude of z is $|z| = e^{\sigma T}$ with a phase of $\theta = \omega T = 2\pi f / F_s$, where F_s is the sampling frequency. To illustrate the mapping from the s-plane to the z-plane, consider the following regions from Figure 4.1.

$\sigma < 0$
Poles on the left side of the $j\omega$ axis (region 2) in the s-plane represent a stable system, and (4.12) yields a magnitude of $|z| < 1$, because $e^{\sigma T} < 1$. As σ varies from $-\infty$ to 0^-, $|z|$ will vary from 0 to 1^-. Hence, poles *inside* the unit circle within region 2 in the z-plane will yield a stable system. The response of such system will be a decaying exponential if the poles are real, or a decaying sinusoid if the poles are complex.

$\sigma > 0$
Poles on the right side of the $j\omega$ axis (region 3) in the s-plane represent an unstable system, and (4.12) yields a magnitude of $|z| > 1$, because $e^{\sigma T} > 1$. As σ varies from 0^+

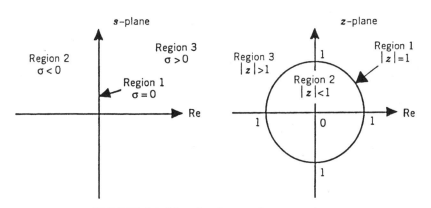

FIGURE 4.1. Mapping from s-plane to z-plane.

to ∞, $|z|$ will vary from 1^+ to ∞. Hence, poles *outside* the unit circle within region 3 in the z-plane will yield an unstable system. The response of such system will be an increasing exponential if the poles are real, or a growing sinusoid if the poles are complex.

$\sigma = 0$

Poles on the $j\omega$ axis (region 1) in the s-plane represent a marginally stable system, and (4.12) yields a magnitude of $|z| = 1$, which corresponds to region 1. Hence, poles *on* the unit circle in region 1 in the z-plane will yield a sinusoid. In Chapter 5 we implement a sinusoidal signal by programming a difference equation with its poles *on* the unit circle. Note that from Exercise 4.2 the poles of $X(s) = \sin n\omega T$ in (4.9) or $X(s) = \cos n\omega T$ in (4.11) are the roots of $z^2 - 2z \cos \omega T + 1$, or

$$p_{1,2} = \frac{2\cos\omega T \pm \sqrt{4\cos^2\omega T - 4}}{2}$$
$$= \cos\omega T \pm \sqrt{-\sin^2\omega T} = \cos\omega T \pm j \sin\omega T \qquad (4.13)$$

The magnitude of each pole is

$$|p_1| = |p_2| = \sqrt{\cos^2\omega T + \sin^2\omega T} = 1 \qquad (4.14)$$

The phase of z is $\theta = \omega T = 2\pi f/F_s$. As the frequency f varies from zero to $\pm F_s/2$, the phase θ will vary from 0 to π.

4.1.2 Difference Equations

A digital filter is represented by a difference equation in a similar fashion as an analog filter is represented by a differential equation. To solve a difference equation, we need to find the z-transform of expressions such as $x(n - k)$, which corresponds to the kth derivative $d^k x(t)/dt^k$ of an analog signal $x(t)$. The order of the difference equation is determined by the largest value of k. For example, $k = 2$ represents a second-order derivative. From (4.5)

$$X(z) = \sum_{n=0}^{\infty} x(n)z^{-n} = x(0) + x(1)z^{-1} + x(2)z^{-2} + \cdots \qquad (4.15)$$

Then the z-transform of $x(n - 1)$, which corresponds to a first-order derivative dx/dt, is

$$ZT[x(n-1)] = \sum_{n=0}^{\infty} x(n-1)z^{-n}$$
$$= x(-1) + x(0)z^{-1} + x(1)z^{-2} + x(2)z^{-3} + \cdots$$
$$= x(-1) + z^{-1}[x(0) + x(1)z^{-1} + x(2)z^{-2} + \cdots]$$
$$= x(-1) + z^{-1}X(z) \qquad (4.16)$$

where we used (4.15), and $x(-1)$ represents the initial condition associated with a first-order difference equation. Similarly, the ZT of $x(n-2)$, equivalent to a second derivative $d^2x(t)/dt^2$ is

$$ZT[x(n-2)] = \sum_{n=0}^{\infty} x(n-2)z^{-n}$$
$$= x(-2) + x(-1)z^{-1} + x(0)z^{-2} + x(1)z^{-3} + \cdots$$
$$= x(-2) + x(-1)z^{-1} + z^{-2}[x(0) + x(1)z^{-1} + \cdots]$$
$$= x(-2) + x(-1)z^{-1} + z^{-2}X(z) \tag{4.17}$$

where $x(-2)$ and $x(-1)$ represent the two initial conditions required to solve a second-order difference equation. In general,

$$ZT[x(n-k)] = z^{-k}\sum_{m=1}^{k} x(-m)z^{m} + z^{k}X(z) \tag{4.18}$$

If the initial conditions are all zero, then $x(-m) = 0$ for $m = 1, 2, \ldots, k$, and (4.18) reduces to

$$ZT[x(n-k)] = z^{-k}X(z) \tag{4.19}$$

4.2 DISCRETE SIGNALS

A discrete signal $x(n)$ can be expressed as

$$x(n) = \sum_{m=-\infty}^{\infty} x(m)\delta(n-m) \tag{4.20}$$

where $\delta(n-m)$ is the impulse sequence $\delta(n)$ delayed by m, which is equal to 1 for $n = m$ and is zero otherwise. It consists of a sequence of values $x(1), x(2), \ldots$, where n is the time, and each sample value of the sequence is taken one sample time apart, determined by the sampling interval or sampling period $T = 1/F_s$.

The signals and systems that we deal with in this book are linear and time-invariant, where both superposition and shift invariance apply. Let an input signal $x(n)$ yield an output response $y(n)$, or $x(n) \rightarrow y(n)$. If $a_1x_1(n) \rightarrow a_1y_1(n)$ and $a_2x_2(n) \rightarrow a_2y_2(n)$, then $a_1x_1(n) + a_2x_2(n) \rightarrow a_1y_1(n) + a_2y_2(n)$, where a_1 and a_2 are constants. This is the superposition property, where an overall output response is the sum of the individual responses to each input. Shift-invariance implies that if the input is delayed by m samples, the output response will also be delayed by m samples, or $x(n-m) \rightarrow y(n-m)$. If the input is a unit impulse $\delta(n)$, the resulting output response is $h(n)$, or $\delta(n) \rightarrow h(n)$, and $h(n)$ is designated as the impulse response. A delayed impulse $\delta(n-m)$ yields the output response $h(n-m)$ by the shift-invariance property.

Furthermore, if this impulse is multiplied by $x(m)$, then $x(m)\delta(n - m) \rightarrow x(m)h(n - m)$. Using (4.20), the response becomes

$$y(n) = \sum_{m=-\infty}^{\infty} x(m)h(n-m) \qquad (4.21)$$

which represents a convolution equation. For a causal system, (4.21) becomes

$$y(n) = \sum_{m=-\infty}^{\infty} x(m)h(n-m) \qquad (4.22)$$

Letting $k = n - m$ in (4.22) yields

$$y(n) = \sum_{k=0}^{\infty} h(k)x(n-k) \qquad (4.23)$$

4.3 FINITE IMPULSE RESPONSE FILTERS

Filtering is one of the most useful signal processing operations [1–47]. Digital signal processors are now available to implement digital filters in real time. The TMS320C6x instruction set and architecture makes it well suited for such filtering operations. An analog filter operates on continuous signals and is typically realized with discrete components such as operational amplifiers, resistors, and capacitors. However, a digital filter, such as a finite impulse response (FIR) filter, operates on discrete-time signals and can be implemented with a digital signal processor such as the TMS320C6x. This involves use of an ADC to capture an external input signal, processing the input samples, and sending the resulting output through a DAC.

Within the last few years, the cost of digital signal processors has been reduced significantly, which adds to the numerous advantages that digital filters have over their analog counterparts. These include higher reliability, accuracy, and less sensitivity to temperature and aging. Stringent magnitude and phase characteristics can be realized with a digital filter. Filter characteristics such as center frequency, bandwidth, and filter type can readily be modified. A number of tools are available to design and implement within a few minutes an FIR filter in real time using the TMS320C6x-based DSK. The filter design consists of the approximation of a transfer function with a resulting set of coefficients.

Different techniques are available for the design of FIR filters, such as a commonly used technique that utilizes the Fourier series, as discussed in the Section 4.4. Computer-aided design techniques such as that of Parks and McClellan are also used for the design of FIR filters [5,6].

The convolution equation (4.23) is very useful for the design of FIR filters, since we can approximate it with a finite number of terms, or

$$y(n) = \sum_{k=0}^{N-1} h(k)x(n-k) \tag{4.24}$$

If the input is a unit impulse $x(n) = \delta(0)$, the output impulse response will be $y(n) = h(n)$. We will see in Section 4.4 how to design an FIR filter with N coefficients $h(0), h(1), \ldots, h(N-1)$, and N input samples $x(n), x(n-1), \ldots, x(n-(N-1))$. The input sample at time n is $x(n)$, and the delayed input samples are $x(n-1), \ldots, x(n-(N-1))$. Equation (4.24) shows that an FIR filter can be implemented with knowledge of the input $x(n)$ at time n and of the delayed inputs $x(n-k)$. It is nonrecursive and no feedback or past outputs are required. Filters with feedback (recursive) that require past outputs are discussed in Chapter 5. Other names used for FIR filters are transversal and tapped-delay filters.

The z-transform of (4.24) with zero initial conditions yields

$$Y(z) = h(0)X(z) + h(1)z^{-1}X(z) + h(2)z^{-2}X(z) + \cdots + h(N-1)z^{-(N-1)}X(z) \tag{4.25}$$

Equation (4.24) represents a convolution in time between the coefficients and the input samples, which is equivalent to a multiplication in the frequency domain, or

$$Y(z) = H(z)X(z) \tag{4.26}$$

where $H(z) = ZT[h(k)]$ is the transfer function, or

$$\begin{aligned} H(z) &= \sum_{k=0}^{N-1} h(k)z^{-k} = h(0) + h(1)z^{-1} + h(2)z^{-2} + \cdots + h(N-1)z^{-(N-1)} \\ &= \frac{h(0)z^{(N-1)} + h(1)z^{N-2} + h(2)z^{N-3} + \cdots + h(N-1)}{z^{N-1}} \end{aligned} \tag{4.27}$$

which shows that there are $N-1$ poles, all of which are located at the origin. Hence, this FIR filter is inherently stable, with its poles located only inside the unit circle. We usually describe an FIR filter as a filter with "no poles." Figure 4.2 shows an FIR filter structure representing (4.24) and (4.25).

A very useful feature of an FIR filter is that it can guarantee *linear phase*. The linear phase feature can be very useful in applications such as speech analysis, where phase distortion can be very critical. For example, with linear phase, all input sinusoidal components are delayed by the same amount. Otherwise, harmonic distortion can occur.

The Fourier transform of a delayed input sample $x(n-k)$ is $e^{-j\omega kT}X(j\omega)$, yielding a phase of $\theta = -\omega kT$, which is a linear function in terms of ω. Note that the group delay function, defined as the derivative of the phase, is a constant, or $d\theta/d\omega = -kT$.

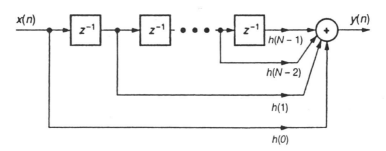

FIGURE 4.2. FIR filter structure showing delays.

4.4 FIR IMPLEMENTATION USING FOURIER SERIES

The design of an FIR filter using a Fourier series method is such that the magnitude response of its transfer function $H(z)$ approximates a desired magnitude response. The transfer function desired is

$$H_d(\omega) = \sum_{n=-\infty}^{\infty} C_n e^{jn\omega T} \qquad |n| < \infty \tag{4.28}$$

where C_n are the Fourier series coefficients. Using a normalized frequency variable v such that $v = f/F_N$, where F_N is the Nyquist frequency, or $F_N = F_s/2$, the desired transfer function in (4.28) can be written as

$$H_d(v) = \sum_{n=-\infty}^{\infty} C_n e^{jn\pi v} \tag{4.29}$$

where $\omega T = 2\pi f/F_s = \pi v$ and $|v| < 1$. The coefficients C_n are defined as

$$C_n = \frac{1}{2}\int_{-1}^{1} H_d(v)e^{-jn\pi v}dv$$
$$= \frac{1}{2}\int_{-1}^{1} H_d(v)(\cos n\pi v - j\sin n\pi v)dv \tag{4.30}$$

Assume that $H_d(v)$ is an even function (frequency selective filter); then (4.30) reduces to

$$C_n = \int_{0}^{1} H_d(v)\cos n\pi v\, dv \qquad n \geq 0 \tag{4.31}$$

since $H_d(v)\sin n\pi v$ is an odd function and

$$\int_{-1}^{1} H_d(v)\sin n\pi v\, dv = 0$$

with $C_n = C_{-n}$. The desired transfer function $H_d(v)$ in (4.29) is expressed in terms of an infinite number of coefficients, and to obtain a realizable filter, we must truncate (4.29), which yields the approximated transfer function

$$H_a(v) = \sum_{n=-Q}^{Q} C_n e^{jn\pi v} \qquad (4.32)$$

where Q is positive and finite and determines the order of the filter. The larger the value of Q, the higher the order of the FIR filter and the better the approximation in (4.32) of the desired transfer function. The truncation of the infinite series with a finite number of terms results in ignoring the contribution of the terms outside a rectangular window function between $-Q$ and $+Q$. In Section 4.5 we see how the characteristics of a filter can be improved by using window functions other than rectangular.

Let $z = e^{j\pi v}$; then (4.32) becomes

$$H_a(z) = \sum_{n=-Q}^{Q} C_n z^n \qquad (4.33)$$

with the impulse response coefficients $C_{-Q}, C_{-Q+1}, \ldots, C_{-1}, C_0, C_1, \ldots, C_{Q-1}, C_Q$. The approximated transfer function in (4.33), with positive powers of z, implies a noncausal or not realizable filter that would produce an output before an input is applied. To remedy this situation, we introduce a delay of Q samples in (4.33) to yield

$$H(z) = z^{-Q} H_a(z) = \sum_{n=-Q}^{Q} C_n z^{n-Q} \qquad (4.34)$$

Let $n - Q = -i$; then $H(z)$ in (4.34) becomes

$$H(z) = \sum_{i=0}^{2Q} C_{Q-i} z^{-i} \qquad (4.35)$$

Let $h_i = C_{Q-i}$ and $N - 1 = 2Q$; then $H(z)$ becomes

$$H(z) = \sum_{i=0}^{N-1} h_i z^{-i} \qquad (4.36)$$

where $H(z)$ is expressed in terms of the impulse response coefficients h_i, and $h_0 = C_Q, h_1 = C_{Q-1}, \ldots, h_Q = C_0, h_{Q+1} = C_{-1} = C_1, \ldots, h_{2Q} = C_{-Q}$. The impulse response coefficients are symmetric about h_Q, with $C_n = C_{-n}$.

The order of the filter is $N = 2Q + 1$. For example, if $Q = 5$, the filter will have 11 coefficients $h_0, h_1, \ldots, h_{10}$, or

$$h_0 = h_{10} = C_5$$
$$h_1 = h_9 = C_4$$
$$h_2 = h_8 = C_3$$
$$h_3 = h_7 = C_2$$
$$h_4 = h_6 = C_1$$
$$h_5 = C_0$$

Figure 4.3 shows the desired transfer functions $H_d(v)$ ideally represented for the frequency-selective filters: lowpass, highpass, bandpass, and bandstop for which the coefficients $C_n = C_{-n}$ can be found.

1. *Lowpass:* $C_0 = v_1$

$$C_n = \int_0^{v_1} H_d(v) \cos n\pi v \, dv = \frac{\sin n\pi v_1}{n\pi} \qquad (4.37)$$

2. *Highpass:* $C_0 = 1 - v_1$

$$C_n = \sum_{v_1}^{1} H_d(v) \cos n\pi v \, dv = -\frac{\sin n\pi v_1}{n\pi} \qquad (4.38)$$

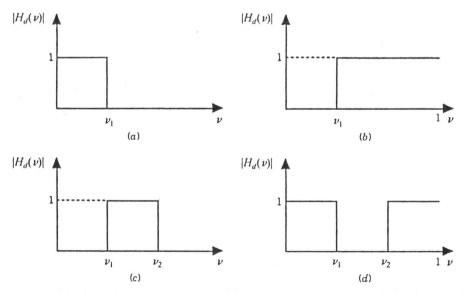

FIGURE 4.3. Desired transfer function: (*a*) lowpass; (*b*) highpass; (*c*) bandpass; (*d*) bandstop.

3. *Bandpass:* $C_0 = v_2 - v_1$

$$C_n = \int_{v_1}^{v_2} H_d(v)\cos n\pi v \, dv = \frac{\sin n\pi v_2 - \sin n\pi v_1}{n\pi} \tag{4.39}$$

4. *Bandstop:* $C_0 = 1 - (v_2 - v_1)$

$$C_n = \int_0^{v_1} H_d(v)\cos n\pi v \, dv + \int_{v_2}^1 H_d(v)\cos n\pi v \, dv = \frac{\sin n\pi v_1 - \sin n\pi v_2}{n\pi} \tag{4.40}$$

where v_1 and v_2 are the normalized cutoff frequencies shown in Figure 4.3. Several filter-design packages are currently available for the design of FIR filters, as discussed later. When we implement an FIR filter, we develop a generic program such that the specific coefficients will determine the filter type (e.g., whether lowpass or bandpass).

Exercise 4.3: Lowpass FIR Filter

We will find the impulse response coefficients of an FIR filter with $N = 11$, a sampling frequency of 10 kHz, and a cutoff frequency $f_c = 1$ kHz. From (4.37),

$$C_0 = v_1 = \frac{f_c}{F_N} = 0.2$$

where $F_N = F_s/2$ is the Nyquist frequency and

$$C_n = \frac{\sin 0.2 n\pi}{n\pi} \qquad n = \pm 1, \pm 2, \dots, \pm 5 \tag{4.41}$$

Since the impulse response coefficients $h_i = C_{Q-i}$, $C_n = C_{-n}$, and $Q = 5$, the impulse response coefficients are

$$
\begin{aligned}
h_0 &= h_{10} = 0 & h_3 &= h_7 = 0.1514 \\
h_1 &= h_9 = 0.0468 & h_4 &= h_6 = 0.1872 \\
h_2 &= h_8 = 0.1009 & h_5 &= 0.2
\end{aligned}
\tag{4.42}
$$

These coefficients can be calculated with a utility program (on the accompanying disk) and inserted within a generic filter program, as described later. Note the symmetry of these coefficients about $Q = 5$. While $N = 11$ for an FIR filter is low for a practical design, doubling this number can yield an FIR filter with much better characteristics, such as selectivity. For an FIR filter to have linear phase, the coefficients must be symmetric, as in (4.42).

4.5 WINDOW FUNCTIONS

We truncated the infinite series in the transfer function equation (4.29) to arrive at (4.32). We essentially put a rectangular window function with an amplitude of 1 between $-Q$ and $+Q$ and ignored the coefficients outside that window. The wider this rectangular window, the larger Q is and the more terms we use in (4.32) to get a better approximation of (4.29). The rectangular window function can therefore be defined as

$$w_R(n) = \begin{cases} 1 & \text{for } |n| \leq Q \\ 0 & \text{otherwise} \end{cases} \tag{4.43}$$

The transform of the rectangular window function $\omega_R(n)$ yields a sinc function in the frequency domain. It can be shown that

$$W_R(v) = \sum_{n=-Q}^{Q} e^{jn\pi v} = e^{-jQ\pi v}\left(\sum_{n=0}^{2Q} e^{jn\pi v}\right) = \frac{\sin\left[\left(\dfrac{2Q+1}{2}\right)\pi v\right]}{\sin(\pi v/2)} \tag{4.44}$$

which is a sinc function that exhibits high sidelobes or oscillations caused by the abrupt truncation, specifically, near discontinuities.

A number of window functions are currently available to reduce these high-amplitude oscillations; they provide a more gradual truncation to the infinite series expansion. However, while these alternative window functions reduce the amplitude of the sidelobes, they also have a wider mainlobe, which results in a filter with lower selectivity. A measure of a filter's performance is a ripple factor that compares the peak of the first sidelobe to the peak of the main lobe (their ratio). A compromise or trade-off is to select a window function that can reduce the sidelobes while approaching the selectivity that can be achieved with the rectangular window function. The width of the mainlobe can be reduced by increasing the width of the window (order of the filter). We will later plot the magnitude response of an FIR filter that shows the undesirable sidelobes.

In general, the Fourier series coefficients can be written as

$$C_n' = C_n w(n) \tag{4.45}$$

where $w(n)$ is the window function. In the case of the rectangular window function, $C_n' = C_n$. The transfer function in (4.36) can then be written as

$$H'(z) = \sum_{i=0}^{N-1} h_i' z^{-i} \tag{4.46}$$

where

$$h_i' = C_{Q-i}' \qquad 0 \le i \le 2Q \tag{4.47}$$

The rectangular window has its highest sidelobe level, down by only −13 dB from the peak of its mainlobe, resulting in oscillations with an amplitude of considerable size. On the other hand, it has the narrowest mainlobe that can provide high selectivity. The following window functions are commonly used in the design of FIR filters [12].

4.5.1 Hamming Window

The Hamming window function [12,25] is

$$w_H(n) = \begin{cases} 0.54 + 0.46\cos(n\pi/Q) & \text{for } |n| \le Q \\ 0 & \text{otherwise} \end{cases} \tag{4.48}$$

which has the highest or first sidelobe level at approximately −43 dB from the peak of the main lobe.

4.5.2 Hanning Window

The Hanning or raised cosine window function is

$$w_{HA}(n) = \begin{cases} 0.5 + 0.5\cos(n\pi/Q) & \text{for } |n| \le Q \\ 0 & \text{otherwise} \end{cases} \tag{4.49}$$

which has the highest or first sidelobe level at approximately −31 dB from the peak of the mainlobe.

4.5.3 Blackman Window

The Blackman window function is

$$w_B(n) = \begin{cases} 0.42 + 0.5\cos(n\pi/Q) + 0.08\cos(2n\pi/Q) & |n| \le Q \\ 0 & \text{otherwise} \end{cases} \tag{4.50}$$

which has the highest sidelobe level down to approximately −58 dB from the peak of the mainlobe. While the Blackman window produces the largest reduction in the sidelobe compared with the previous window functions, it has the widest mainlobe. As with the previous windows, the width of the mainlobe can be decreased by increasing the width of the window.

4.5.4 Kaiser Window

The design of FIR filters with the Kaiser window has become very popular in recent years. It has a variable parameter to control the size of the sidelobe with respect to the mainlobe. The Kaiser window function is

$$
w_K(n) = \begin{cases} I_0(b)/I_0(a) & |n| \leq Q \\ 0 & \text{otherwise} \end{cases}
\tag{4.51}
$$

where a is an empirically determined variable, and $b = a[1 - (n/Q)^2]^{1/2}$. $I_0(x)$ is the modified Bessel function of the first kind defined by

$$
I_0(x) = 1 + \frac{0.25x^2}{(1!)^2} + \frac{(0.25x^2)^2}{(2!)^2} + \cdots = 1 + \sum_{n=1}^{\infty} \left[\frac{(x/2)^n}{n!} \right]^2
\tag{4.52}
$$

which converges rapidly. A trade-off between the size of the sidelobe and the width of the mainlobe can be achieved by changing the length of the window and the parameter a.

4.5.5 Computer-Aided Approximation

An efficient technique is the computer-aided iterative design based on the Remez exchange algorithm, which produces equiripple approximation of FIR filters [5,6]. The order of the filter and the edges of both passbands and stopbands are fixed, and the coefficients are varied to provide this equiripple approximation. This minimizes the ripple in both the passbands and the stopbands. The transition regions are left unconstrained and are considered as "don't care" regions, where the solution may fail. Several commercial filter design packages include the Parks–McClellan algorithm for the design of an FIR filter.

4.6 PROGRAMMING EXAMPLES USING C AND ASM CODE

Within minutes, an FIR filter can be designed and implemented in real time. Several filter design packages are available for the design of FIR filters. They are described in Appendix D using MATLAB [48] and in Appendix E using DigiFilter and a home-made package (on the accompanying disk).

Several examples illustrate the implementation of FIR filters. Most of the programs are in C. A few examples using mixed C and ASM code illustrate the use of a circular buffer as a more efficient way to update delay samples, with the circular buffer in internal or external memory. The convolution equation (4.24) is used to program and implement these filters, or

$$y(n) = \sum_{i=0}^{N-1} h(i)x(n-i)$$

We can arrange the impulse response coefficients within a buffer (array) so that the first coefficient, $h(0)$, is at the beginning (first location) of the buffer (lower-memory address). The last coefficient, $h(N-1)$, can reside at the end (last location) of the coefficients buffer (higher-memory address). The delay samples are organized in memory so that the newest sample, $x(n)$, is at the beginning of the samples buffer, while the oldest sample, $x(n-(N-1))$, is at the end of the buffer. The coefficients and the samples can be arranged in memory as shown in Table 4.1. Initially, all the samples are set to zero.

Time n

The newest sample, $x(n)$, at time n is acquired from an ADC and stored at the beginning of the sample buffer. The filter's output at time n is computed from the convolution equation (4.24), or

$$y(n) = h(0)x(n) + h(1)x(n-1) + \cdots + h(N-2)x(n-(N-2)) \\ + h(N-1)x(n-(N-1))$$

The delay samples are then updated, so that $x(n-k) = x(n+1-k)$ can be used to calculate $y(n+1)$, the output for the next unit of time, or sample period T_s. All the samples are updated except the newest sample. For example, $x(n-1) = x(n)$, and $x(n-(N-1)) = x(n-(N-2))$. This updating process has the effect of "moving the data" (down) in memory (see Table 4.2, associated with time $n+1$).

Time n + 1

At time $n+1$, a new input sample $x(n+1)$ is acquired and stored at the top of the sample buffer, as shown in Table 4.2. The output $y(n+1)$ can now be calculated as

TABLE 4.1   Memory Organization for Coefficients and Samples (Initially)

i	Coefficients	Samples
0	h(0)	x(n)
1	h(1)	x(n - 1)
2	h(2)	x(n - 2)
.	.	.
.	.	.
.	.	.
N - 1	h(N - 1)	x(n - (N - 1))

TABLE 4.2 Memory Organization to Illustrate Update of Samples

	Samples		
Coefficients	Time n	Time n + 1	Time n + 2
h(0)	x(n)	x(n + 1)	x(n + 2)
h(1)	x(n - 1)	x(n)	x(n + 1)
h(2)	x(n - 2)	x(n - 1)	x(n)
.	.	.	.
.	.	.	.
.	.	.	.
h(N - 3)	x(n - (N - 3))	x(n - (N - 4))	x(n - (N - 5))
h(N - 2)	x(n - (N - 2))	x(n - (N - 3))	x(n - (N - 4))
h(N - 1)	x(n - (N - 1))	x(n - (N - 2))	x(n - (N - 3))

$$y(n + 1) = h(0)x(n + 1) + h(1)x(n) + \cdots + h(N - 2)x(n - (N - 3))$$
$$+ h(N - 1)x(n - (N - 2))$$

The samples are then updated for the next unit of time.

Time n + 2

At time $n + 2$, a new input sample, $x(n + 2)$, is acquired. The output becomes

$$y(n + 2) = h(0)x(n + 2) + h(1)x(n + 1) + \cdots + h(N - 1)x(n - (N - 3))$$

This process continues to calculate the filter's output and updating the delay samples at each unit of time (sample period).

Example 4.8 illustrates four different ways of arranging the coefficients and samples in memory and of calculating the convolution equation (e.g., the newest sample at the end of the buffer and the oldest sample at the beginning).

Example 4.1: FIR Filter Implementation: Bandstop and Bandpass (FIR)

Figure 4.4 shows a listing of the C source program $FIR.c$, which implements an FIR filter. It is a generic FIR program, since the coefficient file included, $bs2700.cof$ (Figure 4.5), specifies the filter's characteristics. This coefficient file, which contains 89 coefficients, represents an FIR bandstop (notch) filter centered at 2700 Hz. The number of coefficients N is defined in the coefficient file. This filter was designed using MATLAB's graphical user interface (GUI) filter designer SPTOOL, described in Appendix D. Figure 4.6 shows the filter's characteristics (MATLAB's order of 88 corresponds to 89 coefficients).

A buffer $dly[N]$ is created for the delay samples. The newest input sample, $x(n)$, is acquired through $dly[0]$ and stored at the beginning of the buffer. The

```
//Fir.c FIR filter. Include coefficient file with length N

#include "bs2700.cof"              //coefficient file BS @ 2700Hz
int yn = 0;                        //initialize filter's output
short dly[N];                      //delay samples

interrupt void c_int11()          //ISR
{
     short i;

     dly[0] = input_sample();     //newest input @ top of buffer
     yn = 0;                      //initialize filter's output
     for (i = 0; i< N; i++)
        yn += (h[i] * dly[i]);    //y(n) += h(i)* x(n-i)
     for (i = N-1; i > 0; i--)    //starting @ bottom of buffer
        dly[i] = dly[i-1];        //update delays with data move

     output_sample(yn >> 15);     //output filter
     return;
}

void main()
   {
     comm_intr();                 //init DSK, codec, McBSP
     while(1);                    //infinite loop
   }
```

FIGURE 4.4. Generic FIR program (FIR.c).

```
//BS2700.cof FIR bandstop coefficients designed with MATLAB

#define N 89                        //number of coefficients

short h[N]={-14,23,-9,-6,0,8,16,-58,50,44,-147,119,67,-245,
           200,72,-312,257,53,-299,239,20,-165,88,0,105,
           -236,33,490,-740,158,932,-1380,392,1348,-2070,
           724,1650,-2690,1104,1776,-3122,1458,1704,29491,
           1704,1458,-3122,1776,1104,-2690,1650,724,-2070,
           1348,392,-1380,932,158,-740,490,33,-236,105,0,
           88,-165,20,239,-299,53,257,-312,72,200,-245,67,
           119,-147,44,50,-58,16,8,0,-6,-9,23,-14};
```

FIGURE 4.5. Coefficients for a FIR bandstop filter (bs2700.cof).

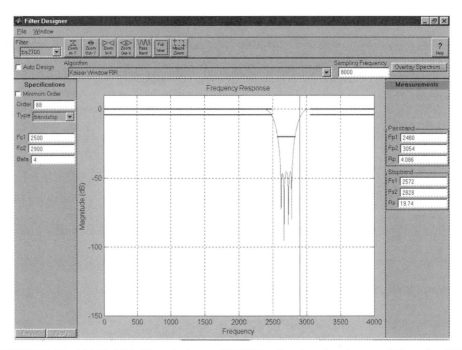

FIGURE 4.6. MATLAB's filter designer SPTOOL, showing the characteristics of a FIR bandstop filter centered at 2700 Hz.

coefficients are stored in another buffer, $h[N]$, with $h[0]$ at the beginning of the coefficients' buffer. The samples and coefficients are then arranged in their respective buffer, as shown in Table 4.1.

Two "for" loops are used within the interrupt service routine (we will also implement an FIR filter using one loop). The first loop implements the convolution equation with N coefficients and N delay samples, for a specific time n. At time n the output is

$$y(n) = h(0)x(n) + h(1)x(n-1) + \cdots + h(N-1)x(n-(N-1))$$

The delay samples are then updated within the second loop to be used for calculating $y(n)$ at time $n+1$, or $y(n+1)$. The newly acquired input sample always resides at the beginning of the samples buffer (in this example). The memory location that contained the sample $x(n)$ now contains the newly acquired sample $x(n+1)$. The output $y(n+1)$ at time $n+1$ is then calculated. This scheme uses a data move to update the delay samples.

Example 4.8 illustrates how various memory organizations can be used for both the delay samples and the filter coefficients, as well as updating the delay samples within the same loop as the convolution equation. We also illustrate the use of a circular buffer with a pointer to update the delay samples, in lieu of moving the data in memory. The output is scaled (right-shifted by 15) before it is sent to the codec's DAC. This allows for a fixed-point implementation as well.

Bandstop, Centered at 2700 Hz (bs2700.cof)

Build and run this project as **FIR**. Input a sinusoidal signal and vary the input frequency slightly below and above 2700 Hz. Verify that the output is a minimum at 2700 Hz.

Figure 4.7 shows a plot of CCS project windows. It shows the FFT magnitude of the filter's coefficients h (see Example 1.3, using a starting address of h) using a 128-point FFT. The characteristics of the FIR bandstop filter, centered at 2700 Hz, are displayed. It also shows a CCS time-domain plot, or the impulse response of the filter.

With noise as input, the output frequency response of the bandpass filter can also be verified. The pseudorandom noise sequence developed in Chapter 2, or another noise source (see Appendix D), can be used as input to the FIR filter, as illustrated later. Figure 4.8 shows a plot of the frequency response of the filter with a notch at 2700 Hz implemented in real time. This plot is obtained using an HP 3561A dynamic signal analyzer with an input noise source from the analyzer. The roll-off at approximately 3500 Hz is due to the antialiasing lowpass filter on the codec.

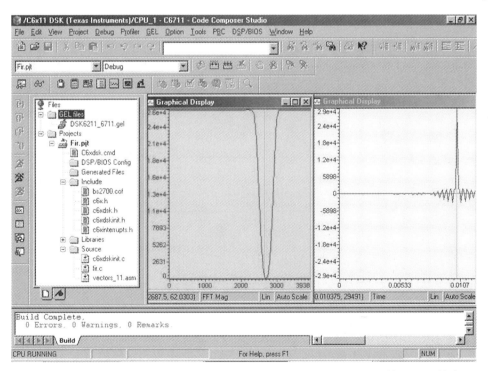

FIGURE 4.7. CCS plots displaying the FFT magnitude of the bandstop filter's coefficients and its impulse response.

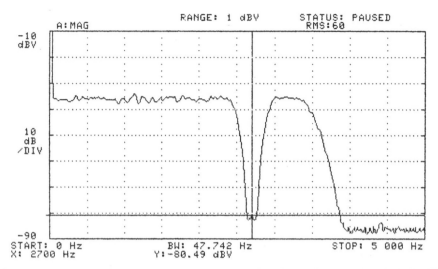

FIGURE 4.8. Output frequency response of FIR bandstop filter centered at 2700 Hz, obtained with a signal analyzer.

Bandpass, Centered at 1750 Hz (bp1750.cof)

Within CCS, edit the program FIR.c to include the coefficient file *bp1750.cof* in lieu of *bs2700.cof*. The file *bp1750.cof* represents an FIR bandpass filter (81 coefficients) centered at 1750 Hz, as shown in Figure 4.9. This filter was designed

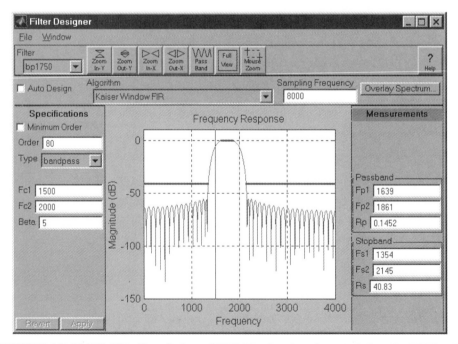

FIGURE 4.9. MATLAB's filter designer SPTOOL, showing characteristics of a FIR bandpass filter centered at 1750 Hz.

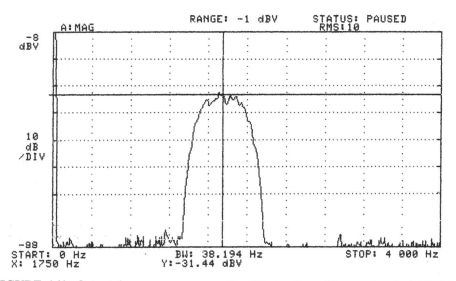

FIGURE 4.10. Output frequency response of a FIR bandpass filter centered at 1750 Hz, obtained with a signal analyzer.

with MATLAB's SPTOOL (Appendix D). Select the incremental Build and the new coefficient file *bp1750.cof* will automatically be included in the project. Run again and verify an FIR bandpass filter centered at 1750 Hz. Figure 4.10 shows a real-time plot of the output frequency response obtained with the HP signal analyzer.

Example 4.2: Effects on Voice Using Three FIR Lowpass Filters (*FIR3LP*)

Figure 4.11 shows a listing of the program *FIR3lp.c*, which implements three FIR lowpass filters with cutoff frequencies at 600, 1500, and 3000 Hz, respectively. The three lowpass filters were designed with MATLAB's SPTOOL to yield the corresponding three sets of coefficients. This example expands on the generic FIR program in Example 4.1.

LP_number selects the desired lowpass filter to be implemented. For example, if LP_number is set to 1, h[1][i] is equal to hlp600[i] (within the "for" loop in the function *main*), which is the address of the first set of coefficients. The coefficients file LP600.cof represents an 81-coefficient FIR lowpass filter with a 600-Hz cutoff frequency, using the Kaiser window function. Figure 4.12 shows a listing of this coefficient file (the other two sets are on the disk). That filter is then implemented. LP_number can be changed to 2 or 3 to implement the 1500- or 3000-Hz lowpass filter, respectively. With the GEL file FIR3LP.gel (Figure 4.13), one can vary LP_number from 1 to 3 and slide through the three different filters.

```
//FIR3LP.c FIR using three lowpass coefficients with three different BW

#include "lp600.cof"                        //coeff file LP @ 600 Hz
#include "lp1500.cof"                       //coeff file LP @ 1500 Hz
#include "lp3000.cof"                       //coeff file LP @ 3000 Hz
short LP_number = 1;                        //start with 1st LP filter
int   yn = 0;                              //initialize filter's output
short dly[N];                              //delay samples
short h[3][N];                             //filter characteristics 3xN

interrupt void c_int11()                   //ISR
{
    short i;

    dly[0] = input_sample();               //newest input @ top of buffer
    yn = 0;                                //initialize filter output
    for (i = 0; i< N; i++)
       yn +=(h[LP_number][i]*dly[i]);      //y(n) += h(LP#,i)*x(n-i)
    for (i = N-1; i > 0; i--)              //starting @ bottom of buffer
       dly[i] = dly[i-1];                  //update delays with data move
    output_sample(yn >> 15);               //output filter
    return;                                //return from interrupt
}

void main()
{
    short i;

    for (i=0; i<N; i++)
       {
          dly[i] = 0;                      //init buffer
          h[1][i] = hlp600[i];             //start addr of LP600 coeff
          h[2][i] = hlp1500[i];            //start addr of LP1500 coeff
          h[3][i] = hlp3000[i];            //start addr of LP3000 coeff
       }
    comm_intr();                           //init DSK, codec, McBSP
    while(1);                              //infinite loop
}
```

FIGURE 4.11. FIR program to implement three different FIR lowpass filters using a slider for selection (FIR3LP.c).

Build this project as **FIR3LP**. Use the .wav file *TheForce.wav* (on the disk) as input (see Appendix D) and observe the effects of the three lowpass filters on the input voice. With the lower bandwidth of 600 Hz, using the first set of coefficients, the frequency components of the speech signal above 600 Hz are suppressed. Connect the output to a speaker or a spectrum analyzer to verify such results, and observe the different bandwidths of the three FIR lowpass filters.

```
//LP600.cof FIR lowpass filter coefficients using Kaizer window

#define N 81          //length of filter

short hlp600[N] = {0,-6,-14,-22,-26,-24,-13,8,34,61,80,83,63,19,-43,-113,
-171,-201,-185,-117,0,146,292,398,428,355,174,-99,-416,-712,-905,-921,
-700,-218,511,1424,2425,3391,4196,4729,4915,4729,4196,3391,2425,1424,
511,-218,-700,-921,-905,-712,-416,-99,174,355,428,398,292,146,0,-117,
-185,-201,-171,-113,-43,19,63,83,80,61,34,8,-13,-24,-26,-22,-14,-6,0};
```

FIGURE 4.12. Coefficient file for a FIR lowpass filter with 600-Hz cutoff frequency (LP600.cof).

```
/*FIR3LP.gel Gel file to step through 3 different LP filters*/

menuitem "Filter Characteristics"

slider Filter(1,3,1,1,filterparameter)   /*from 1 to 3,incr by 1*/
{
      LP_number = filterparameter;       /*for 3 LP filters*/
}
```

FIGURE 4.13. GEL file for selecting one of three FIR lowpass filter coefficients (FIR3LP.gel).

Example 4.3: Implementation of Four Different Filters: Lowpass, Highpass, Bandpass, and Bandstop (FIR4types)

This example is similar to Example 4.2 and illustrates the gel (slider) file to step through four different types of FIR filters. Each filter has 81 coefficients, designed with MATLAB's SPTOOL. The four coefficient files (on the accompanying disk) are:

1. *lp1500.cof*: lowpass with bandwidth of 1500 Hz
2. *hp2200.cof*: highpass with bandwidth of 2200 Hz
3. *bp1750.cof*: bandpass with center frequency at 1750 Hz
4. *bs790.cof*: bandstop with center frequency at 790 Hz

The program *FIR4types.c* (on disk) implements this project. The program FIR3LP.c (Example 4.2) is modified slightly to incorporate a fourth filter.

Build and run this project as **FIR4types**. Load the GEL file *FIR4types.gel* (on the disk) and verify the implementation of the four different FIR filters. This example can readily be expanded to implement more FIR filters.

Figure 4.9 shows the characteristics of the FIR bandpass filter centered at 1750 Hz obtained with MATLAB's filter designer; and Figure 4.10 shows its frequency response obtained with an HP signal analyzer.

```
//FIRPRN.c FIR with internally generated input noise sequence

#include "bp55.cof"                        //BP @ Fs/4 coeff file in float
#include "noise_gen.h"                      //header file for noise sequence
int dly[N];                                 //delay samples
short fb;                                    //feedback variable
shift_reg sreg;

short prn(void)                             //pseudorandom noise generation
{
 short prnseq;                              //for pseudorandom sequence

 if(sreg.bt.b0)                             //sequence {1,-1}
     prnseq = -16000;                       //scaled negative noise level
 else
     prnseq = 16000;                        //scaled positive noise level
 fb =(sreg.bt.b0)^(sreg.bt.b1);            //XOR bits 0,1
 fb ^=(sreg.bt.b11)^(sreg.bt.b13);         //with bits 11,13 ->fb
 sreg.regval<<=1;                           //shift register 1 bit to left
 sreg.bt.b0 = fb;                           //close feedback path

 return prnseq;                             //return sequence
}

interrupt void c_int11()                    //ISR
{
     int i;
     int yn = 0;                            //initialize filter's output

     dly[0] = prn();                        //input noise sequence
     for (i = 0; i< N; i++)
        yn +=(h[i]*dly[i]);                 //y(n)+= h(i)*x(n-i)
     for (i = N-1; i > 0; i--)              //start @ bottom of buffer
        dly[i] = dly[i-1];                  //data move to update delays

     output_sample(yn);                     //output filter
     return;                                //return from interrupt
}

void main()
{
     short i;

     sreg.regval = 0xFFFF;                  //shift register to nominal values
     fb = 1;                                //initial feedback value
     for (i = 0; i<N; i++)
        dly[i] = 0;                         //init buffer
     comm_intr();                           //init DSK, codec, McBSP
     while(1);                              //infinite loop
}
```

FIGURE 4.14. FIR program with pseudorandom noise sequence as input (FIRPRN.c).

Example 4.4: FIR Implementation with Pseudorandom Noise Sequence as Input to Filter (FIRPRN)

The program *FIRPRN.c* (Figure 4.14) implements an FIR filter using an internally generated pseudorandom noise as input to the filter. This input is the pseudorandom noise sequence generated in Example 2.16. The coefficient file *BP55.cof* uses a float data format and is shown in Figure 4.15. [A filter development package (on disk) that generates filter coefficients in float or hexadecimal format is described in Appendix E.] It represents a 55-coefficient FIR bandpass filter with a center frequency at $F_s/4$.

Build this project as **FIRPRN**. Run this project and verify that the output is an FIR bandpass filter centered at 2 kHz. To verify the output as the noise sequence, output `dly[0]` in lieu of `yn` when calling the function `output_sample`.

Testing Different FIR Filters

Halt the program. Edit the C source program to include and test different coefficient files (on the disk) that represent different FIR filters, all using float format values. Each coefficient file contains 55 coefficients (except `comb14.cof`).

1. `BS55.cof`: bandstop with center frequency $F_s/4$
2. `BP55.cof`: bandpass with center frequency $F_s/4$
3. `LP55.cof`: lowpass with cutoff frequency $F_s/4$
4. `HP55.cof`: highpass with bandwidth $F_s/4$
5. `Pass2b.cof`: with two passbands
6. `Pass3b.cof`: with three passbands

```
//bp55.cof Coefficients for bandpass FIR filter centered @ Fs/4

#define N 55                    //number of coefficients

float h[N]=
{1.7619E-017,  7.0567E-003,  2.2150E-018,-1.0962E-002,  4.0310E-017,
 1.3946E-002,  7.1787E-018,-1.4588E-002,  3.9928E-017,  1.1474E-002,
 5.9881E-018,-3.5159E-003,-6.6174E-018,-9.7476E-003,-1.7919E-017,
 2.7932E-002,-9.4329E-017,-4.9740E-002,  3.3834E-017,  7.3066E-002,
-3.6228E-017,-9.5284E-002,  3.2194E-017,  1.1365E-001,-2.2165E-017,
-1.2576E-001,  7.8980E-018,  1.3000E-001,  7.8980E-018,-1.2576E-001,
-2.2165E-017,  1.1365E-001,  3.2194E-017,-9.5284E-002,-3.6228E-017,
 7.3066E-002,  3.3834E-017,-4.9740E-002,-9.4329E-017,  2.7932E-002,
-1.7919E-017,-9.7476E-003,-6.6174E-018,-3.5159E-003,  5.9881E-018,
 1.1474E-002,  3.9928E-017,-1.4588E-002,  7.1787E-018,  1.3946E-002,
 4.0310E-017,-1.0962E-002,  2.2150E-018,  7.0567E-003,  1.7619E-017};
```

FIGURE 4.15. Coefficient file in float format for a FIR bandpass filter centered at $F_s/4$ (BP55.cof).

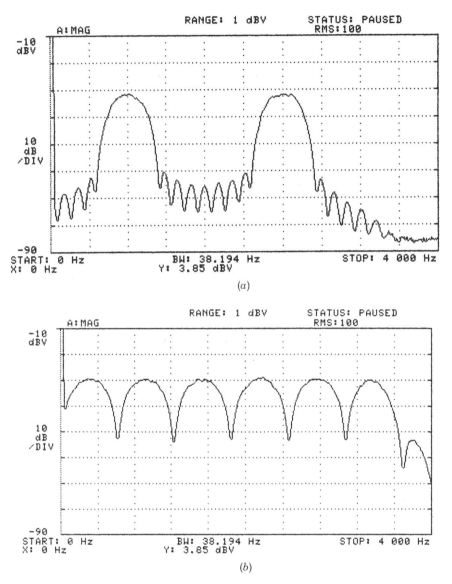

FIGURE 4.16. Output frequency responses obtained with HP analyzer: (*a*) FIR filter with two passbands; (*b*) FIR comb filter.

7. `Pass4b.cof`: with four passbands
8. `Stop3b.cof`: with three stopbands
9. `Comb14.cof`: with multiple notches (comb filter)

Figure 4.16*a* shows the real-time output frequency response of an FIR filter with two passbands, using the coefficient file *pass2b.cof*. This filter was designed with MATLAB. Figure 4.16*b* shows the frequency response of the comb filter, using the coefficients file *comb14.cof*. These plots were obtained with the HP 3561A signal analyzer.

Example 4.5: FIR Filter with Frequency Response Plot
Using CCS (FIRbuf)

Figure 4.17 shows a listing of the program FIRbuf.c, which implements an FIR filter and stores the filter's output into a buffer. The FFT magnitude of the filter's output frequency response can then be plotted using CCS. Example 4.1 illustrated the implementation of an FIR filter using a generic program that includes the coefficient file representing the characteristics of a desired filter. Example 1.2 shows how one can store the output into a buffer so that it can be plotted within CCS. The program FIRbuf.c is based on these two earlier examples. The coefficient file bp41.cof represents a 41-coefficient FIR bandpass filter centered at 1 kHz. The output buffer has a size of 1024.

Build this project as **FIRbuf**. Verify that the output is a bandpass filter, centered at 1 kHz. Halt the processor.

With noise as input to the filter, the output frequency response can be plotted using CCS. The shareware utility Goldwave generates different signals, including noise, using a sound card (see Appendix E). The output from the sound card with the noise generated by Goldwave can be used as the input to the DSK.

Select View → Graph → Time/Frequency. Select/set for:

1. *Display type:* FFT magnitude
2. *Start address:* yn_buffer
3. *Acquisition buffer size:* 1024
4. *FFT frame size:* 1024
5. *FFT order:* 10
6. *DSP data type:* 16-bit signed integer
7. *Sampling rate:* 8000 Hz

Use default for the other fields. The FFT order is M, where 2^M = FFT frame size. Run the program and verify the output frequency response of the filter plotted in Figure 4.18.

Example 4.6: FIR Filter with Internally Generated Pseudorandom Noise as Input to Filter and Output Stored in Memory (FIRPRNbuf)

This example builds on Example 2.16, noise_gen, which generates a pseudorandom noise sequence, and Example 4.5, FIRbuf, which implements an FIR filter with the filter's output also stored in a memory buffer. Figure 4.19 shows a listing of the program FIRPRNbuf.c, which implements this project example.

The input to the filter is a software-generated noise sequence using dly[0] as the newest noise sequence. The coefficient file BP41.cof, which represents a 41-coefficient FIR bandpass filter, is the same as that used in Example 4.5.

```
//Firbuf.c FIR filter with output in buffer plotted with CCS

#include "bp41.cof"                    //BP @ 1 kHz coefficient file

int yn = 0;                           //initialize filter's output
short dly[N];                         //delay samples
short buffercount = 0;                //init buffer count
const short bufferlength = 1024;      //buffer size
short yn_buffer[1024];                //output buffer

interrupt void c_int11()              //ISR
{
    short i;

    dly[0] = input_sample();          //newest input @ top of buffer
    yn = 0;                           //initialize filter's output
    for (i = 0; i< N; i++)
       yn +=(h[i]*dly[i]) >> 15;      //y(n)+=h(i)*x(n-i)
    for (i = N-1; i > 0; i--)         //start @ bottom of buffer
       dly[i] = dly[i-1];             //data move to update delays

    output_sample(yn);                //output filter

    yn_buffer[buffercount] = yn;      //filter's output into buffer
    buffercount++;                    //increment buffer count
    if(buffercount==bufferlength)     //if buffer count = size
       buffercount = 0;               //reinitialize buffer count
    return;                           //return from interrupt
}

void main()
{
    comm_intr();                      //init DSK, codec, McBSP
    while(1);                         //infinite loop
}
```

FIGURE 4.17. FIR program with the filter output stored in memory (FIRbuf.c).

Build and run this project as **FIRPRNbuf**. Verify the output frequency response of a 1-kHz FIR bandpass filter. Goldwave can also be used as a crude spectrum analyzer to obtain the frequency response of the filter (with the output of the DSK connected to the input of the sound card).

Using CCS, verify the FFT magnitude plot as shown in Figure 4.20, using 1024 points. The address of the output buffer is yn_buffer. Figure 4.21 shows the frequency response of the FIR bandpass filter, centered at $F_s/8$, displayed using an HP analyzer.

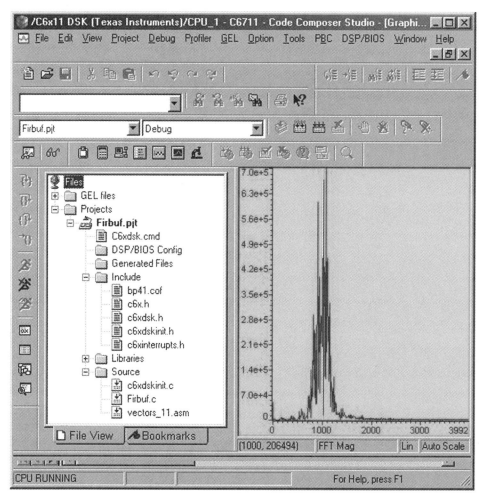

FIGURE 4.18. Output frequency response of a 1-kHz FIR bandpass filter plotted with CCS using external noise as input for project `FIRbuf`.

Change the output buffer so that the noise sequence is stored in memory using

```
yn_buffer[i] = dly[0];
```

Run the program again and plot the FFT magnitude of the noise sequence. It is not quite flat since the resulting plot is not averaged.

You can also output the noise sequence using

```
output_sample(dly[0]);
```

in the program. With the output to a spectrum analyzer with averaging capability, verify that the noise spectrum is quite flat until about 3500 Hz, the bandwidth of the antialiasing filter on the codec (looks like a lowpass filter with a bandwidth of

```
//FIRPRNbuf.c  FIR filter with input noise sequence & output in buffer

#include "bp41.cof"                 //BP @ 1 kHz coefficient file
#include "noise_gen.h"              //header file for noise sequence
int yn = 0;                         //initialize filter's output
short dly[N];                       //delay samples
short buffercount = 0;              //init buffer count
const short bufferlength = 1024;    //buffer size
short yn_buffer[1024];              //output buffer
short fb;                           //feedback variable
shift_reg sreg;

short prn(void)                     //pseudorandom noise generation
{
 short prnseq;                      //for pseudorandom sequence

 if(sreg.bt.b0)                     //sequence {1,-1}
     prnseq = -8000;               //scaled negative noise level
 else
     prnseq = 8000;                //scaled positive noise level
 fb =(sreg.bt.b0)^(sreg.bt.b1);     //XOR bits 0,1
 fb ^=(sreg.bt.b11)^(sreg.bt.b13);  //with bits 11,13 ->fb
 sreg.regval<<=1;                   //shift register 1 bit to left
 sreg.bt.b0 = fb;                   //close feedback path

  return prnseq;
}

interrupt void c_int11()            //ISR
{
     short i;

     dly[0] = prn();               //input noise sequence
     yn = 0;                       //initialize filter's output
     for (i = 0; i< N; i++)
        yn +=(h[i]*dly[i]) >>15;    //y(n)+=h(i)*x(n-i)
     for (i = N-1; i > 0; i--)      //start @ bottom of buffer
        dly[i] = dly[i-1];          //data move to update delays

     output_sample(yn);            //output filter

     yn_buffer[buffercount] = yn;  //filter's output into buffer
     buffercount++;                 //increment buffer count
     if(buffercount==bufferlength)  //if buffer count = size
        buffercount = 0;            //reinitialize buffer count
     return;                        //return from interrupt
}

void main()
{
     sreg.regval = 0xFFFF;         //shift register to nominal values
     fb = 1;                        //initial feedback value
     comm_intr();                   //init DSK, codec, McBSP
     while(1);                      //infinite loop
}
```

FIGURE 4.19. FIR program with an input pseudorandom noise sequence and output stored in the memory buffer (FIRPRNbuf.c).

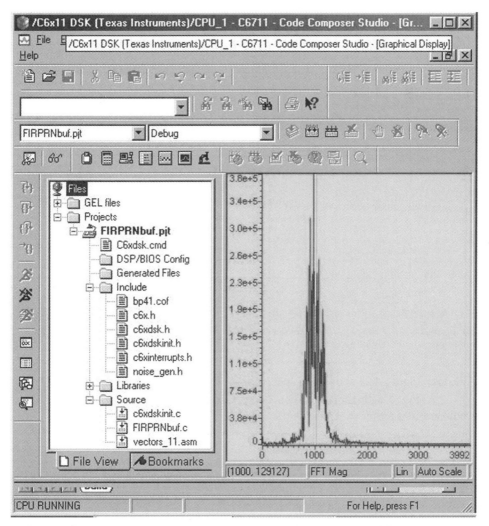

FIGURE 4.20. CCS output frequency response of a 1-kHz FIR bandpass filter using an internally generated noise sequence as input to the filter for project FIRPRNbuf.

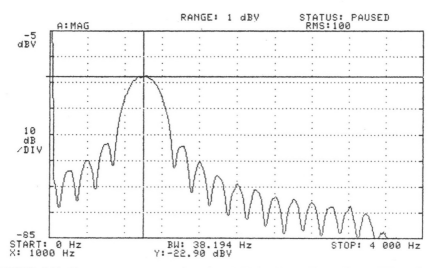

FIGURE 4.21. Frequency response of a 1-kHz FIR bandpass filter using an HP analyzer.

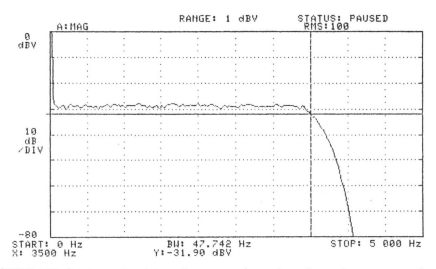

FIGURE 4.22. Spectrum of an internally generated pseudorandom noise sequence using an HP analyzer.

3500 Hz). Figure 4.22 shows the spectrum of this noise sequence using the HP analyzer (averaged with the analyzer). Use a GEL file to develop a slider so that the DSK output is either the noise sequence generated internally, `dly[0]`, or the filter's output `y(n)`.

Example 4.7: Two Notch Filters to Recover Corrupted Input Voice (*NOTCH2*)

This example illustrates the implementation of two notch (bandstop) FIR filters to remove two undesired sinusoidal signals corrupting an input voice signal. The voice signal (*TheForce.wav*, on the disk) was ADDed (using Goldwave) with the two undesired sinusoidal signals at frequencies of 900 Hz and 2700 Hz, to produce the corrupted input signal *corruptvoice.wav* (on the disk).

Figure 4.23 shows a listing of the program *NOTCH2.c*, which implements the two notch filters in cascade (series). Two coefficient files, BS900.cof and BS2700.cof (on the disk), each containing 89 coefficients and designed with MATLAB, are included in the filter program *NOTCH2.c*. They represent two FIR notch filters, centered at 900 Hz and 2700 Hz, respectively. A buffer is used for the delay samples of each filter. The output of the first notch filter, centered at 900 Hz, becomes the input to the second notch filter, centered at 2700 Hz.

Build this project as **NOTCH2**. Input (play) the corrupted voice signal *corruptvoice.wav*. Verify that the slider in position 1 (as set initially) outputs the corrupted voice signal, as shown in Figure 4.24. This plot is obtained with Goldwave using the DSK output as the input to a sound card (see Appendix E). The plot is shown on only one side (left channel) since a mono signal is used. Observe the two spikes (representing the two sinusoidal signals) at 900 Hz and 2700 Hz, respectively.

```
//Notch2.C Two FIR notch filters to remove two sinusoidal noise signals

#include "BS900.cof"              //BS @ 900 Hz coefficient file
#include "BS2700.cof"             //BS @ 2700 Hz coefficient file
short dly1[N]={0};                //delay samples for 1st filter
short dly2[N]={0};                //delay samples for 2nd filter
int y1out = 0, y2out = 0;         //init output of each filter
short out_type = 1;              //slider for output type

interrupt void c_int11()          //ISR
{
     short i;

     dly1[0] = input_sample();    //newest input @ top of buffer
     y1out = 0;                   //init output of 1st filter
     y2out = 0;                   //init output of 2nd filter
     for (i = 0; i< N; i++)
        y1out += h900[i]*dly1[i]; //y1(n)+=h900(i)*x(n-i)

     dly2[0]=(y1out >>15);        //out of 1st filter->in 2nd filter
     for (i = 0; i< N; i++)
        y2out += h2700[i]*dly2[i]; //y2(n)+=h2700(i)*x(n-i)

     for (i = N-1; i > 0; i--)     //from bottom of buffer
       {
         dly1[i] = dly1[i-1];      //update samples of 1st buffer
         dly2[i] = dly2[i-1];      //update samples of 2nd buffer
       }

     if (out_type==1)             //if slider is in position 1
     output_sample(dly1[0]);      //corrupted input(voice+sines)
     if (out_type==2)
     output_sample(y2out>>15);    //output of 2nd filter (voice)
     return;                      //return from ISR
}

void main()
{
     comm_intr();                 //init DSK, codec, McBSP
     while(1);                    //infinite loop
}
```

FIGURE 4.23. Program with two FIR notch filters in cascade to remove two undesired sinusoidal signals (NOTCH2.c).

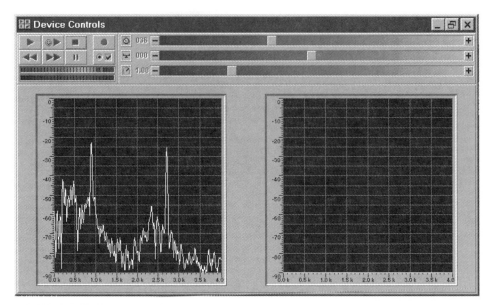

FIGURE 4.24. Spectrum of voice signal corrupted by two sinusoidal signals at 900 and 2700 Hz (obtained with Goldwave).

Change the slider to position 2 and verify that the two undesirable sinusoidal signals are removed.

Also output *y1out* through the function *output_sample* (rebuild) and verify that only the 2700-Hz corrupts the input voice signal.

Example 4.8: FIR Implementation Using Four Different Methods (FIR4ways)

Figure 4.25 shows a listing of the program *FIR4ways.c*, which implements an FIR filter using four alternative methods for convolving/updating the delay samples. This example extends Example 4.1, where the first method (method A) is used. In this first method with two "for" loops, the delay samples are arranged in memory with the newest sample at the beginning of the buffer and the oldest sample at the end of the buffer. The convolution starts with the newest sample and the first coefficient using

$$y(n) = h(0)x(n) + h(1)x(n-1) + \cdots + h(N-1)x(n-(N-1))$$

Each data value is "moved down" in memory to update the delay samples, with the newest sample being the newly acquired input sample. The size of the array for the delay samples is now set at $N + 1$ and not N, to illustrate the third method (method C). The other three methods use a buffer size of N for the delay samples. The bottom (end) of the buffer in this example refers to memory location N, not $N + 1$. Note

```
//FIR4ways.c FIR with alternative ways of storing/updating samples

#include "bp41.cof"                 //BP @ 1 kHz coefficient file
#define METHOD 'D'                  //change to B or C or D
int yn = 0;                         //initialize filter's output
short dly[N+1];                     //delay samples array(one extra)

interrupt void c_int11()            //ISR
{
     short i;
     yn = 0;                        //initialize filter's output

#if METHOD == 'A'                   //if 1st method
     dly[0] = input_sample();       //newest sample @ top of buffer
     for (i = 0; i< N; i++)
        yn += (h[i] * dly[i]);      //y(n)=h[0]*x[n]+..+h[N-1]x[n-(N-1)]
     for (i = N-1; i > 0; i--)      //from bottom of buffer
        dly[i] = dly[i-1];          //update sample data move "down"

#elif METHOD == 'B'                 //if 2nd method
     dly[0] = input_sample();       //newest sample @ top of buffer
     for (i = N-1; i >= 0; i--)     //start @ bottom to convolve
       {
        yn += (h[i] * dly[i]);      //y=h[N-1]x[n-(N-1)]+...+h[0]x[n]
        dly[i] = dly[i-1];          //update sample data move "down"
       }

#elif METHOD == 'C'                 //use xtra memory location
     dly[0] = input_sample();       //newest sample @ top of buffer
     for (i = N-1; i>=0; i--)       //start @ bottom of buffer
       {
        yn += (h[i] * dly[i]);      //y=h[N-1]x[n-(N-1)]+...+h[0]x[n]
        dly[i+1] = dly[i];          //update sample data move "down"
       }

#elif METHOD == 'D'                 //1st convolve before loop
     dly[N-1] = input_sample();     //newest sample @ bottom of buffer
     yn = h[N-1] * dly[0];          //y=h[N-1]x[n-(N-1)] (only one)
     for (i = 1; i<N; i++)          //convolve the rest
       {
        yn +=(h[N-(i+1)]*dly[i]);   //h[N-2]x[n-(N-2)]+...+h[0]x[n]
        dly[i-1] = dly[i];          //update sample data move "up"
       }
#endif
output_sample(yn >> 15);            //output filter
return;                             //return from ISR
}

void main()
{
     comm_intr();                   //init DSK, codec, McBSP
     while(1);                      //infinite loop
}
```

FIGURE 4.25. FIR program using four alternative methods for convolution and updating of delay samples (FIR4ways.c).

that in this case the unused data $x(n - N)$ in memory location $(N + 1)$ is not updated, by using the index $i < N$.

The second method (method B) performs the convolution and updates the delay samples using one loop. The convolution starts with the oldest coefficient and the oldest sample, "moving up" through the buffers using

$$y(n) = h(N - 1)x(n - (N - 1)) + h(N - 2)x(n - (N - 2)) + \cdots + h(0)x(n)$$

The updating scheme is similar to the first method. In method B, when $i = 0$, the newest sample is updated by an invalid data value residing at the memory location preceding the start of the sample buffer. But this invalid data item is then replaced by a newly acquired input sample with `dly[0]` before calculating $y(n)$ for the next unit of time. Or, one could use an "if" statement to update the delay samples for all values of i except for $i = 0$.

The third method uses $N + 1$ memory locations to update the delay samples. The unused data at memory location $N + 1$ is also updated. The extra memory location is used so that a valid data item in that location is not overwritten.

The fourth method performs the first convolution expression "outside" the loop. The delay samples in the previous methods were arranged in memory so that the newest sample, $x(n)$, is at the beginning of the buffer and the oldest sample, $x(n - (N - 1))$, is at the end. However, in this method, the newest input sample is acquired through `dly[N - 1]` so that the newest sample is now at the end of the buffer and the updating process moves the data up.

Build and run this project as **FIR4ways**. Verify that the output is an FIR bandpass filter centered at 1 kHz. Change the method to test (define) the other three methods and verify that the resulting output is the same.

Example 4.9: Voice Scrambler Using Filtering and Modulation (Scram16k)

This example illustrates a voice scrambling/descrambling scheme. The approach makes use of basic algorithms for filtering and modulation. Modulation was introduced in Example 2.14. With voice as input, the resulting output is scrambled voice. The original unscrambled voice is recovered when the output of the DSK is used as the input to a second DSK running the same program.

An up-sampling scheme is used to process at a sampling rate of 16 kHz in lieu of the 8-kHz rate set with the AD535 codec. This results in a better performance, allowing for a wider input signal bandwidth.

The scrambling method used is commonly referred to as *frequency inversion*. It takes an audio range, represented by the band 0.3 to 3 kHz, and "folds" it about a carrier signal. The frequency inversion is achieved by multiplying (modulating) the audio input by a carrier signal, causing a shift in the frequency spectrum with upper and lower sidebands. On the lower sideband that represents the audible speech range, the low tones are high tones, and vice versa.

Figure 4.26 is a block diagram of the scrambling scheme. At point A we have a

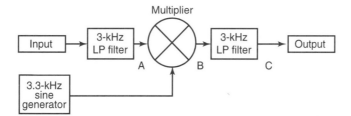

FIGURE 4.26. Block diagram of scrambler/descrambler scheme.

bandlimited signal 0 to 3 kHz. At point B we have a double-sideband signal with suppressed carrier. At point C the upper sideband is filtered out. Its attractiveness comes from its simplicity, since only simple DSP algorithms are utilized: filtering, sine generation/modulation, and up-sampling (due to low sampling frequency with the AD535 codec).

Figure 4.27 shows a listing of the program `Scram16k.c`, which implements this project. The input signal is first lowpass filtered and the resulting output (at point A) is multiplied (modulated) by a 3.3-kHz sine function with data values in a buffer (lookup table). The modulated signal (at point B) is filtered again, and the overall output is a scrambled signal (at point C).

There are three functions in Figure 4.27 in addition to the function *main*. One of the functions, *filtmodfilt*, calls a filter function to implement the first lowpass filter as an antialiasing filter. The resulting output (filtered input) becomes the input to a multiplier/modulator. The function *sinemod* modulates (multiplies) the filtered input with the 3.3-kHz sine data values. This produces higher and lower sideband components. The modulated output is again filtered, so that only the lower sideband components are kept.

The up-sampling scheme to obtain a 16-kHz sampling rate is achieved by "processing" the data twice and retaining only the second result. This allows for a wider input signal bandwidth to be scrambled.

A buffer is used to store the 114 coefficients that represent the lowpass filter. The coefficient file *lp114.cof* is on disk. Two other buffers are used for the delay samples, one for each filter. The samples are arranged in memory as

$$x(n - (N - 1)), x(n - (N - 2)), \ldots, x(n - 1), x(n)$$

with the oldest sample at the beginning of the buffer and the newest sample at the end (bottom) of the buffer. The file *sine160.h* with 160 data values over 33 cycles is on disk. The frequency generated is $f = F_s$ (number of cycles)/(number of points) $= 16,000(33)/160 = 3.3$ kHz.

Using the resulting output as the input to a second DSK running the same algorithm, the original unscrambled input is recovered as the output of the second DSK. Note that the program can still run on the first DSK when it is disconnected from the parallel port cable (DB25 cable).

Build and run this project as **Scram16k**. First test this project using a 2-kHz input sine wave. The resulting output is a lower sideband signal of 1.3 kHz, obtained as

```
//Scram16k.c Voice scrambler/de-scrambler program

#include "sine160.h"                 //sine data values
#include "LP114.cof"                 //filter coefficient file
short filtmodfilt(short data);
short filter(short inp,short *dly);
short sinemod(short input);
static short filter1[N],filter2[N];
short input, output;

void main()
{
 short i;

 comm_poll();                        //init DSK using polling
 for (i=0; i< N; i++)
   {
    filter1[i] = 0;                  //init 1st filter buffer
    filter2[i] = 0;                  //init 2nd filter buffer
   }
 while(1)
   {
    input=input_sample();            //input new sample data
    filtmodfilt(input);              //process sample twice(upsample)
    output=filtmodfilt(input);       //and throw away 1st result
    output_sample(output);           //then output
   }
}

short filtmodfilt(short data)        //filtering & modulating
{
 data = filter(data,filter1);        //newest in ->1st filter
 data = sinemod(data);               //modulate with 1st filter out
 data = filter(data,filter2);        //2nd LP filter
 return data;
}

short filter(short inp,short *dly)   //implements FIR
{
 short i;
 int yn;

 dly[N-1] = inp;                     //newest sample @bottom buffer
 yn = dly[0] * h[N-1];               //y(0)=x(n-(N-1))*h(N-1)
 for (i = 1; i < N; i++)             //loop for the rest
   {
    yn += dly[i] * h[N-(i+1)];       //y(n)=x[n-(N-1-i)]*h[N-1-i]
    dly[i-1] = dly[i];               //data up to update delays
   }
 yn = ((yn) >>15);                   //filter's output
 return yn;                          //return y(n) at time n
}

short sinemod(short input)           //sine generation/modulation
{
 static short i=0;
 input=(input*sine160[i++])>>11;     //(input)*(sine data)
 if(i>= NSINE) i = 0;                //if end of sine table
 return input;                       //return modulated signal
}
```

FIGURE 4.27. Voice scrambler program (Scram16k.c).

(3.3 kHz – 2 kHz). The upper sideband signal of (3.3 kHz + 2 kHz) is filtered out by the second lowpass filter.

A second DSK is used to recover/unscramble the original signal (simulating the receiving end). Use the output of the first DSK as the input to the second DSK. Run the same program on the second DSK. This produces the reverse procedure, yielding the original unscrambled signal. If the same 2-kHz original input is considered, the 1.3 kHz as the scrambled signal becomes the input to the second DSK. The resulting output is the original signal of 2 kHz (3.3 kHz – 1.3 kHz), the lower sideband signal.

With a sweeping input sinusoidal signal increasing in frequency, the resulting output is the sweeping signal "decreasing" in frequency. Use as input the .wav file *TheForce.wav* and verify the scrambling/descrambling scheme.

Interception of the speech signal can be made more difficult by changing the modulation frequency dynamically and including (or omitting) the carrier frequency according to a predefined sequence: for example, a code for no modulation, another for modulating at frequency f_{c1}, and a third code for modulating at frequency f_{c2}.

This project was first implemented using the TMS320C25 [49] and also on the TMS320C31 DSK without the need for up-sampling.

Example 4.10: Illustration of Aliasing Effects with Down-Sampling (aliasing)

Figure 4.28 shows a listing of the program aliasing.c, which implements this project. To illustrate the effects of aliasing, the processing rate is down-sampled by a factor of 2, to an equivalent 4-kHz rate. Note that the antialiasing and reconstruction filters on the AD535 codec are fixed and connot be bypassed or altered. Up-sampling and lowpass filtering are needed to output the 4-kHz rate samples to the AD535 codec sampling at 8 kHz.

Build this project as **aliasing**. Load the slider file aliasing.gel (on the disk). With antialiasing initially set to zero in the program, aliasing will occur.

1. Input a sinusoidal signal and verify that for an input signal frequency up to 2 kHz, the output is essentially a loop program (delayed input). Increase the input signal frequency to 2.5 kHz and verify that the output is an aliased 1.5-kHz signal. Similarly, a 3- and a 3.5-kHz input signal yield an aliased output signal of 1 and 0.5 kHz, respectively. Input signals with frequencies beyond 3.5 kHz are supressed due to the AD535 codec's antialiasing filter.

2. Change the slider position to 1, so that antialiasing at the down-sampled rate of 4 kHz is desired. For an input signal frequency up to about 1.8 kHz, the output is a delayed version of the input. Increase the input signal frequency beyond 1.8 kHz and verify that the output reduces to zero. This is due to the 1.8-kHz antialiasing lowpass filter, implemented using the coefficient file lp33.cof (on the disk).

//Aliasing.c illustration of downsampling, aliasing, upsampling

```
#include "lp33.cof"                    //lowpass at 1.8 kHz
short flag = 0;                        //toggles for 2x down-sampling
float indly[N],outdly[N];             //antialias and reconst delay lines
short i;                               //index
float yn;                              //filter output
short antialiasing = 0;               //init for no antialiasing filter

interrupt void c_int11()              //ISR
{

 indly[0]=(float)(input_sample());    //new sample to antialias filter
 yn = 0.0;                            //initialize downsampled value
 if (flag == 0)                       //discard input sample value
    flag = 1;                         //don't discard at next sampling
 else
  {
   if (antialiasing == 1)            //if antialiasing filter desired
    {                                //compute downsampled value
     for (i = 0 ; i < N ; i++)       //using LP @ 1.8 kHz filter coeffs
        yn += (h[i]*indly[i]);       //filter is implemented using float
    }
   else                             //if filter is bypassed
       yn = indly[0];               //downsampled value is input value
   flag = 0;                        //next input value will be discarded
  }
  for (i = N-1; i > 0; i--)
      indly[i] = indly[i-1];        //update input buffer

  outdly[0] = (yn);                 //input to reconst filter
  yn = 0.0;                         //4 kHz sample values and zeros
  for (i = 0 ; i < N ; i++)         //are filtered at 8 kHz rate
      yn += (h[i]*outdly[i]);       //by reconstruction lowpass filter

  for (i = N-1; i > 0; i--)
      outdly[i] = outdly[i-1];      //update delays

  output_sample((short)(yn));       //8 kHz rate sample
  return;                           //return from interrupt
}

void main()
{
  comm_intr();                      //init DSK, codec, McBSP
  while(1);                         //infinite loop
}
```

FIGURE 4.28. Program to illustrate aliasing and antialiasing down-sampling to a rate of 4kHz (aliasing.c).

In lieu of using a sinusoidal signal as input, you can play `sweep.wav` from Gold-wave (see Appendix E).

Example 4.11: Implementation of an Inverse FIR Filter (`FIRinverse`)

Figure 4.29 shows a listing of the program `FIRinverse.c`, which implements an inverse FIR filter. An original input sequence to an FIR filter can be recovered using an inverse FIR filter. The transfer function of an FIR filter of order N is

```
//FIRinverse.c Implementation of inverse FIR Filter

#include "bp41.cof"              //coefficient file BP @ Fs/8
int yn;                          //filter's output
short dly[N];                    //delay samples
int out_type = 1;                //output type for slider

interrupt void c_int11()         //ISR
{
 short i;

 dly[0] = input_sample();        //newest input sample data
 yn = 0;                         //initialize filter's output

 for (i = 0; i<N; i++)
   yn += (h[i]*dly[i]);          //y(n)+=h(i)*x(n-i)
 if(out_type==1)                 //if slider in position 1
   output_sample(dly[0]);        //original input
 if(out_type==2)
   output_sample(yn>>15);        //output of FIR filter
 if(out_type==3)                 //calculate inverse FIR
   {
   for (i = N-1; i>1; i--)
     yn -= (h[i]*dly[i]);        //calculate inverse FIR filter
   yn = yn/h[0];                 //scale output of inverse filter
   output_sample(yn>>8);         //send output of inverse filter
   }
 for (i = N-1; i>0; i--)         //from bottom of buffer
   dly[i] = dly[i-1];            //update delay samples
 return;                         //return from ISR
}

void main()
{
 comm_intr();                    //init DSK, codec, McBSP
 while(1);                       //infinite loop
}
```

FIGURE 4.29. Program to implement an inverse FIR filter (`FIRinverse.c`).

$$H(z) = \sum_{i=0}^{N-1} h_i z^{-i}$$

where h_i represents the impulse response coefficients. The output sequence of the FIR filter is

$$y(n) = \sum_{i=0}^{N-1} h_i x(n-i) = h_0 x(n) + h_1 x(n-1) + \cdots + h_{N-1} x(n-(N-1))$$

where $x(n-i)$ represents the input sequence. The original input sequence, x, can then be recovered, using $\hat{x}(n)$ as an estimate of $x(n)$, or

$$\hat{x}(n) = \frac{y(n) - \sum_{i=1}^{N-1} h_i \hat{x}(n-i)}{h_0}$$

Build this project as **FIRinverse**. Use noise as input (from Goldwave or from a noise generator, or modify the program to use the pseudorandom noise sequence, etc.). Verify that the output is the input noise sequence, with the slider in position 1 (default). Change the slider to position 2 and verify the output as an FIR bandpass filter centered at 1 kHz. With the slider in position 3, the inverse of the FIR filter is calculated, so that the output is the original input noise sequence.

Example 4.12: FIR Implementation Using C Calling ASM Function (FIRcasm)

The C program *FIRcasm.c* (Figure 4.30) calls the ASM function *FIRcasm-func.asm* (Figure 4.31), which implements an FIR filter.

Build and run this project as **FIRcasm**. Verify that the output is a 1-kHz FIR bandpass filter. Two buffers are created: dly for the data samples and h for the filter's coefficients. On each interrupt, a new data sample is acquired and stored at the end (higher-memory address) of the buffer dly. The delay samples and the filter coefficients are arranged in memory as shown in Table 4.3. The delay samples are stored in memory starting with the oldest sample with the newest sample at the end of the buffer. The coefficients are arranged in memory with $h(0)$ at the beginning of the coefficient buffer and $h(N-1)$ at the end.

The addresses of the delay sample buffer, the filter coefficient buffer, and the size of each buffer are passed to the ASM function through registers A4, B4, and A6, respectively. The size of each buffer through register A6 is doubled since data in each memory location are stored as byte. The pointers A4 and B4 are incremented or decremented every two bytes (two memory locations). The end address of the coefficients' buffer is in B4, which is at $2N-1$.

```
//FIRCASM.c FIR C program calling ASM function fircasmfunc.asm

#include "bp41.cof"              //BP @ Fs/8 coefficient file
int yn = 0;                     //initialize filter's output
short dly[N];                   //delay samples

interrupt void c_int11()        //ISR
{
    dly[N-1] = input_sample();  //newest sample @bottom buffer
    yn = fircasmfunc(dly,h,N);  //to ASM func through A4,B4,A6
    output_sample(yn >> 15);    //filter's output
    return;                     //return from ISR
}

void main()
{
    short i;

    for (i = 0; i<N; i++)
    dly[i] = 0;                 //init buffer for delays
    comm_intr();                //init DSK, codec, McBSP
    while(1);                   //infinite loop
}
```

FIGURE 4.30. C program calling an ASM function for FIR implementation (`FIRcasm.c`).

TABLE 4.3   Memory Organization of Coefficients and Samples for FIRcasm

	Samples	
Coefficients	Time n	Time n + 1
h(0)	A4 → x(n - (N - 1))	A4 → x(n - (N - 2))
h(1)	x(n - (N - 2))	x(n - (N - 3))
h(2)	x(n - (N - 3))	x(n - (N - 4))
.	.	.
.	.	.
.	.	.
h(N - 2)	x(n - 1)	x(n)
B4 → h(N - 1)	x(n)	← newest → x(n + 1)

The two LDH instructions load the content in memory pointed at (whose address is specified by) A4 and the content in memory at the address specified in B4. This loads the oldest samples, $x(n - (N - 1))$ and $h(N - 1)$, respectively. A4 is then post-incremented to point at $x(n - (N - 2))$, and B4 is postdecremented to point at $h(N - 2)$. After the first accumulation, the oldest sample is updated. The content in memory at the address specified by A4 is loaded into A7, then stored at the preceding memory location. This is because A4 is postdecremented *without* modifica-

```
;FIRCASMfunc.asm ASM function called from C to implement FIR
;A4 = Samples address, B4 = coeff address, A6 = filter order
;Delays organized as:x(n-(N-1))...x(n);coeff as h[0]...h[N-1]


                .def    _fircasmfunc
_fircasmfunc:                         ;ASM function called from C
                MV      A6,A1         ;setup loop count
                MPY     A6,2,A6       ;since dly buffer data as byte
                ZERO    A8            ;init A8 for accumulation
                ADD     A6,B4,B4      ;since coeff buffer data as byte
                SUB     B4,1,B4       ;B4=bottom coeff array h[N-1]
loop:                                 ;start of FIR loop
                LDH     *A4++,A2      ;A2=x[n-(N-1)+i] i=0,1,...,N-1
                LDH     *B4--,B2      ;B2=h[N-1-i] i=0,1,...,N-1
                NOP     4
                MPY     A2,B2,A6      ;A6=x[n-(N-1)+i]*h[N-1-i]
                NOP
                ADD     A6,A8,A8      ;accumlate in A8

                LDH     *A4,A7        ;A7=x[(n-(N-1)+i+1]update delays
                NOP     4             ;using data move "up"
                STH     A7,*-A4[1]    ;-->x[(n-(N-1)+i] update sample
                SUB     A1,1,A1       ;decrement loop count
        [A1]    B       loop          ;branch to loop if count # 0
                NOP     5

                MV      A8,A4         ;result returned in A4
                B       B3            ;return addr to calling routine
                NOP     5
```

FIGURE 4.31. FIR ASM function called from C (`FIRcasmfunc.asm`).

tion to point at the memory location containing the oldest sample. As a result, the oldest sample, $x(n - (N - 1))$, is replaced (updated) by $x(n - (N - 2))$. The updating of the delay samples is for the next unit of time. As the output at time n is being calculated, the samples are updated or "primed" for time $(n + 1)$. At time n the filter's output is

$$y(n) = h(N - 1)x(n - (N - 1)) + h(N - 2)x(n - (N - 2)) + \cdots + h(1)x(n - 1) + h(0)x(n)$$

The loop is processed 41 times. For each time n, $n + 1$, and $n + 2$ an output value is calculated, with each sample updated for the next unit of time. The newest sample is also updated in this process, with an invalid data value residing at the memory location beyond the end of the buffer. But this is remedied since for each unit of time, the newest sample, acquired through the ADC of the codec, overwrites it.

Accumulation is in A8 and the result, for each unit of time, is moved to A4 to be returned to the calling function. The address of the calling function is in B3.

Viewing Update of Samples in Memory

1. *Select → View → Memory using a 16-bit hex format and a starting address of* dly. The delay samples are within 82 (not 41) memory locations, each location specified with a byte. The coefficients also occupy 82 memory locations, in the buffer h. You can verify the content in the coefficient buffer stored as a 16-bit or half-word value. Right-click on the memory window and deselect "Float in Main Window" for a better display with both source program and memory.

2. *Select → View → Mixed C/ASM.* Place a breakpoint within the function FIRcasmfunc.asm at the move instruction

```
MV     A8,A4
```

(you can either double-click on that line of code, or right-mouse-click to Toggle Breakpoint).

3. *Select → Debug → Animate (introduced in Chapter 1).* Execution halts at the set breakpoint for each unit of time. Observe the bottom memory location of the delay samples. Verify that the newest sample data value is placed at the end of the buffer. This value is then moved up the buffer. Observe after a while that the samples are being updated, with each value in the buffer moving up in memory. You can also observe the register (pointer) A4 incrementing by 2 (two bytes) and B4 decrementing by 2.

Example 4.13: FIR Implementation Using C Calling Faster ASM Function (FIRcasmfast)

The same C calling program, FIRcasm.c, is used in this example as in Example 4.12. It calls the ASM function *Fircasmfunc.asm* (Figure 4.32) within the file *FIRcasmfuncfast* (not *FIRcasmfunc*).

This function executes faster than the function in Example 4.12 by having parallel instructions and rearranging the sequence of instructions. There are two parallel instructions: LDH/LDH and SUB/LDH.

1. The number of NOPs is reduced from 19 to 11.
2. The SUB instruction to decrement the loop count is moved up the program.
3. The sequence of some instructions changed to "fill" some of the NOP slots.

For example, the conditional branch instruction executes *after* the ADD instruction to accumulate in A8, since branching has five delay slots. Additional changes

```
;FIRCASMfuncfast.asm C-called faster function to implement FIR
          .def     _fircasmfunc
_fircasmfunc:                      ;ASM function called from C
          MV       A6,A1          ;setup loop count
          MPY      A6,2,A6        ;since dly buffer data as byte
          ZERO     A8             ;init A8 for accumulation
          ADD      A6,B4,B4       ;since coeff buffer data as byte
          SUB      B4,1,B4        ;B4=bottom coeff array h[N-1]
   loop:                          ;start of FIR loop
          LDH      *A4++,A2       ;A2=x[n-(N-1)+i] i=0,1,...,N-1
   ||     LDH      *B4--,B2       ;B2=h[N-1-i] i=0,1,...,N-1
          SUB      A1,1,A1        ;decrement loop count
   ||     LDH      *A4,A7         ;A7=x[(n-(N-1)+i+1]update delays
          NOP      4
          STH      A7,*-A4[1]     ;-->x[(n-(N-1)+i] update sample
   [A1]   B        loop           ;branch to loop if count # 0
          NOP      2
          MPY      A2,B2,A6       ;A6=x[n-(N-1)+i]*h[N-1-i]
          NOP
          ADD      A6,A8,A8       ;accumlate in A8

          B        B3             ;return addr to calling routine
          MV       A8,A4          ;result returned in A4
          NOP      4
```

FIGURE 4.32. ASM function called from C for faster execution (FIRcasmfunc-fast.asm).

to make it faster would also make it less comprehensible, due to further resequencing of the instructions.

Build this project as **FIRcasmfast**, so that the linker option names the output executable file FIRcasmfast.out. The resulting output is the 1-kHz bandpass filter in Example 4.12.

Example 4.14: FIR Implementation with C Program Calling ASM Function Using Circular Buffer (FIRcirc)

The C program FIRcirc.c (Figure 4.33) calls the ASM function FIRcircfunc.asm (Figure 4.34), which implements an FIR filter using a circular buffer. This example expands Example 4.13. The coefficients within the file bp1750.cof were designed with MATLAB using the Kaiser window and represent a 128-coefficient FIR bandpass filter with a center frequency of 1750 Hz. Figure 4.35 displays the characteristics of this filter, obtained from MATLAB's filter designer SPTOOL (described in Appendix D).

```
//FIRcirc.c C program calling ASM function using circular buffer

#include "bp1750.cof"                  //BP at 1750 Hz coeff file
int yn = 0;                            //init filter's output

interrupt void c_int11()              //ISR
{
 short sample_data;

 sample_data = input_sample();        //newest input sample data
 yn = fircircfunc(sample_data,h,N);   //ASM func passing to A4,B4,A6
 output_sample(yn >> 15);             //filter's output
 return;                              //return to calling function
}

void main()
{
 comm_intr();                         //init DSK, codec, McBSP
 while(1);                            //infinite loop
}
```

FIGURE 4.33. C program calling an ASM function using a circular buffer (FIRcirc.c).

In lieu of moving the data to update the delay samples, a pointer is used. The 16 LSBs of the address mode register (AMR) are set with a value of

```
0x0040 = 0000 0000 0100 0000
```

This selects A7 mode as the circular buffer pointer register. The 16 MSBs of AMR are set with N = 0x0007 to select the block BK0 as a circular buffer. The buffer size is $2^{N+1} = 256$. A circular buffer is used in this example only for the delay samples. It is also possible to use a second circular buffer for the coefficients. For example, using

```
0x0140 = 0000 0001 0100 0000
```

would select two pointers, B4 and A7.

Within a C program, an inline assembly code can be used with the asm statement. For example,

```
asm(" MVK  0x0040,B6")
```

Note the blank space after the first quote so that the instruction does not start in column 1. The circular mode of addressing eliminates the data move to update the delay samples, since the pointer can be moved to achieve the same result faster.

```
;FIRcircfunc.asm ASM function called from C using circular addressing
;A4=newest sample, B4=coefficient address, A6=filter order
;Delay samples organized: x[n-(N-1)]...x[n]; coeff as h(0)...h[N-1]

              .def   _fircircfunc
              .def   last_addr
              .def   delays
              .sect  "circdata"        ;circular data section
              .align 256               ;align delay buffer 256-byte boundary
delays        .space 256               ;init 256-byte buffer with 0's
last_addr     .int   last_addr-1       ;point to bottom of delays buffer
              .text                    ;code section
_fircircfunc:                          ;FIR function using circ addr
              MV     A6,A1             ;setup loop count
              MPY    A6,2,A6           ;since dly buffer data as byte
              ZERO   A8                ;init A8 for accumulation

              ADD    A6,B4,B4          ;since coeff buffer data as bytes
              SUB    B4,1,B4           ;B4=bottom coeff array h[N-1]

              MVKL   0x00070040,B6     ;select A7 as pointer and BK0
              MVKH   0x00070040,B6     ;BK0 for 256 bytes (128 shorts)

              MVC    B6,AMR            ;set address mode register AMR

              MVK    last_addr,A9      ;A9=last circ addr(lower 16 bits)
              MVKH   last_addr,A9      ;last circ addr (higher 16 bits)
              LDW    *A9,A7            ;A7=last circ addr
              NOP    4
              STH    A4,*A7++          ;newest sample-->last address

loop:                                  ;begin FIR loop
              LDH    *A7++,A2          ;A2=x[n-(N-1)+i] i=0,1,...,N-1
        ||    LDH    *B4--,B2          ;B2=h[N-1-i] i=0,1,...,N-1
              SUB    A1,1,A1           ;decrement count

        [A1]  B      loop              ;branch to loop if count # 0
              NOP    2
              MPY    A2,B2,A6          ;A6=x[n-(N-1)+i]*h[N-1+i]
              NOP
              ADD    A6,A8,A8          ;accumulate in A8

              STW    A7,*A9            ;store last circ addr to last_addr
              B      B3                ;return addr to calling routine
              MV     A8,A4             ;result returned in A4
              NOP    4
```

FIGURE 4.34. C-called ASM function using a circular buffer to update samples (FIR-circfunc.asm).

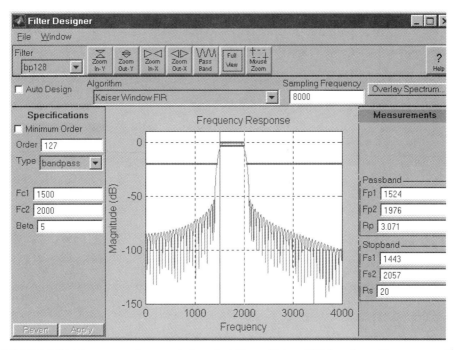

FIGURE 4.35. Frequency characteristics of a 128-coefficient FIR bandpass filter centered at 1750 Hz using MATLAB's filter designer SPTOOL.

Initially, the register pointer A7 points to the last address in the sample buffer. Consider for now the sample buffer only, since it is circular.

1. *Time n.* At time n, A7 points to the end of the buffer, where the newest sample is stored. It is then postincremented to point to the beginning of the buffer, as shown in Table 4.4. Then the section of code within the loop starts, and calculates

$$y(n) = h(N-1)x(n-(N-1)) + h(N-2)x(n-(N-2)) + \cdots$$
$$+ h(1)x(n-1) + h(0)x(n)$$

After the last multiplication, $h(0)x(n)$, A7 is postincremented to point to the beginning address of the buffer. The resulting filter's output at time n is then returned to the calling function. Before the loop starts for each unit of time, A7 always contains the address where the newest sample is to be stored. While the newly acquired sample is passed to the ASM function through A4 at each unit of time $n, n+1, n+2, \ldots$, A4 is stored in A7, which always contains the last address.

2. *Time n + 1.* At time $(n+1)$, the newest sample, $x(n+1)$, is passed to the ASM function through A4. The STH instruction stores that sample into memory

whose address is in A7, which is at the beginning of the buffer. It is then post-incremented to point at the address containing $x(n - (N - 2))$, as shown in Table 4.4. The output is now

$$y(n+1) = h(N-1)x(n-(N-2)) + h(N-2)x(n-(N-3)) + \cdots$$
$$+ h(1)x(n) + h(0)x(n+1)$$

The last multiplication always involves $h(0)$ and the newest sample.

3. *Time $n + 2$.* At time $(n + 2)$, the filter's output is

$$y(n+2) = h(N-1)x(n-(N-3)) + h(N-2)x(n-(N-4)) + \cdots$$
$$+ h(1)x(n+1) + h(0)x(n+2)$$

Note that for each unit of time, the newly acquired sample overwrites the oldest sample at the previous unit of time. At each time $n, n + 1, \ldots$, the filter's output is calculated within the ASM function, and the result is sent to the calling C function, where a new sample is acquired at each sample period.

The conditional branch instruction was moved up as in Example 4.13. Branching to loop takes effect (due to five delay slots) after the ADD instruction to accumulate in A8. One can save the content of AMR at the end of processing one buffer and restore it before using it again with a pair of MVC instructions: MVC AMR, Bx and MVC Bx, AMR using a B register.

Build and run this project as **FIRcirc**. Verify an FIR bandpass filter centered at 1750 Hz. Halt, reset, and reload the program.

Place a breakpoint within the ASM function FIRcircfunc.asm at the branch instruction to return to the calling C function. View memory at the address delays and verify that this buffer of size 256 is initialized to zero. Right-click on the memory

TABLE 4.4   Memory Organization of Coefficients and Samples Using Circular Buffer

Coefficients	Samples		
	Time n	Time n + 1	Time n + 2
h(0)	A7 → x(n - (N - 1))	newest → x(n + 1)	x(n + 1)
h(1)	x(n - (N - 2))	A7 → x(n - (N - 2))	newest → x(n + 2)
h(2)	x(n - (N - 3))	x(n - (N - 3))	A7 → x(n - (N - 3))
.	.	.	.
.	.	.	.
.	.	.	.
h(N - 2)	x(n - 1)	x(n - 1)	x(n - 1)
h(N - 1)	newest → x(n)	x(n)	x(n)

window to toggle "Float in Main Window" (for a better display). Run the program. Execution stops at the breakpoint. Verify that the newest sample (16 bits) is stored at the end (higher address) of the buffer (at 0x3FE and 0x3FF). Memory location 0x400 contains the last address 0x301 where the subsequent sample is to be stored. This address is the beginning of the buffer. View the core registers and verify that A7 contains that address.

Run again and observe the new sample stored at the beginning of the buffer (you can animate now). Note that A7 is incremented to 0x303, 0x305, ... The circular method of updating the delays is more efficient. It is important that the buffer is aligned on a boundary of a power of 2.

Example 4.15: FIR Implementation with C Program Calling ASM Function Using Circular Buffer in External Memory (FIRcirc_ext)

This example implements an FIR filter using a circular buffer in external memory. It expands slightly on Example 4.14. The C program FIRcirc.c in Example 4.14 is modified to obtain *FIRcirc_ext.c* (Figure 4.36) so that it calls the ASM function FIRcircfunc_ext.asm (in lieu of FIRcircfunc.asm).

The linker command file *FIRcirc_ext.cmd* used in this example is listed in

```
//FIRcirc_ext.c  C program calling ASM function using circular buffer

#include "bp1750.cof"                    //BP at 1750 Hz coeff file
int yn = 0;                              //init filter's output

interrupt void c_int11()                 //ISR
{
 short sample_data;

 sample_data = input_sample();           //newest input sample data
 yn = fircircfunc_ext(sample_data,h,N);  //ASM funcn passing to A4,B4,A6
 output_sample(yn >> 15);                //filter's output
 return;                                 //return to calling function
}

void main()
{
 comm_intr();                            //init DSK, codec, McBSP
 while(1);                               //infinite loop
}
```

FIGURE 4.36. C program calling an ASM function with a circular buffer in external memory (FIRcirc_ext.c).

```
;FIRcircfunc_ext.asm Function using circular buffer in external memory
;A4=newest sample, B4=coefficient address, A6=filter order
;Delay samples organized: x[n-(N-1)]...x[n]; coeff as h(0)...h[N-1]

            .def    _fircircfunc_ext
            .def    last_addr
            .def    delays
            .sect   "circdata"      ;circular data section
            .align  256             ;align delay buffer 256-byte boundary
delays      .space  256             ;init 256-byte buffer with 0's
last_addr   .int    last_addr-1
            .text                   ;code section
_fircircfunc_ext:                   ;FIR function using circ addr
            MV      A6,A1           ;setup loop count
            MPY     A6,2,A6         ;since dly buffer data as byte
            ZERO    A8              ;init A8 for accumulation

            ADD     A6,B4,B4        ;since coeff buffer data as bytes
            SUB     B4,1,B4         ;B4=bottom coeff array h[N-1]

            MVKL    0x00070040,B6   ;select A7 as pointer and BK0
            MVKH    0x00070040,B6   ;BK0 for 256 bytes (128 shorts)
            MVC     B6,AMR          ;set address mode register AMR

            MVKL    last_addr,A9    ;A9=bottom circ addr in external mem
            MVKH    last_addr,A9    ;(higher 16 bits)in external circ
            LDW     *A9,A7          ;A7=last circ addr
            NOP     4
            STH     A4,*A7++        ;newest sample-->last address
loop:                               ;begin FIR loop
            LDH     *A7++,A2        ;A2=x[n-(N-1)+i] i-0,1,...,N-1
||          LDH     *B4--,B2        ;B2=h[N-1-i] i=0,1,...,N-1
            SUB     A1,1,A1         ;decrement count

   [A1]     B       loop            ;branch to loop if count # 0
            NOP     2
            MPY     A2,B2,A6        ;A6=x[n-(N-1)+i]*h[N-1+i]
            NOP
            ADD     A6,A8,A8        ;accumulate in A8

            STW     A7,*A9          ;store last circ addr to last_addr
            B       B3              ;return addr to calling routine
            MV      A8,A4           ;result returned in A4
            NOP     4
```

FIGURE 4.37. C-called ASM function with a circular buffer in external memory (FIR-circfunc_ext.asm).

```
/*FIRcirc_ext.cmd Linker file for circular buffer in external memory*/

MEMORY
{
  VECS:        org =          0h, len =       0x220
  IRAM:        org = 0x00000220, len = 0x0000FDC0
  buffer_ext:  org = 0x80000000, len = 0x00000110
  SDRAM:       org = 0x80000110, len = 0x01000000
  FLASH:       org = 0x90000000, len = 0x00020000
}
SECTIONS
{
  circdata :> buffer_ext
  vectors  :> VECS
  .text    :> IRAM
  .bss     :> IRAM
  .cinit   :> IRAM
  .stack   :> IRAM
  .sysmem  :> SDRAM
  .const   :> IRAM
  .switch  :> IRAM
  .far     :> SDRAM
  .cio     :> SDRAM
}
```

FIGURE 4.38. Linker command file for a circular buffer in external memory (FIRcirc_ext.cmd).

Figure 4.38. The section *circdata* designates the memory section buffer_ext, which starts in external memory at 0x80000000.

Build this project as **FIRcirc_ext**. View the memory at the address delays. This should display the external memory section. Verify the circular buffer in external memory. Place a breakpoint as in Example 4.14, animate, and verify that the newest sample is stored at the end of the circular buffer and that the subsequent acquired sample is stored at the beginning of the buffer. Halt, remove the breakpoint, and verify that the output is an FIR bandpass filter centered at 1750 Hz.

REFERENCES

1. W. J. Gomes III and R. Chassaing, Filter design and implementation using the TMS320C6x interfaced with MATLAB, *Proceedings of the 2000 ASEE Annual Conference*, 2000.

2. A. V. Oppenheim and R. Schafer, *Discrete-Time Signal Processing*, Prentice Hall, Upper Saddle River, NJ, 1989.

3. B. Gold and C. M. Rader, *Digital Signal Processing of Signals*, McGraw-Hill, New York, 1969.

4. L. R. Rabiner and B. Gold, *Theory and Application of Digital Signal Processing*, Prentice Hall, Upper Saddle River, NJ, 1975.

5. T. W. Parks and J. H. McClellan, Chebychev approximation for nonrecursive digital filter with linear phase, *IEEE Transactions on Circuit Theory*, Vol. CT-19, 1972, pp. 189–194.

6. J. H. McClellan and T. W. Parks, A unified approach to the design of optimum linear phase digital filters, *IEEE Transactions on Circuit Theory*, Vol. CT-20, 1973, pp. 697–701.

7. J. F. Kaiser, Nonrecursive digital filter design using the I0-sinh window function, *Proceedings of the IEEE International Symposium on Circuits and Systems*, 1974.

8. J. F. Kaiser, Some practical considerations in the realization of linear digital filters, *Proceedings of the 3rd Allerton Conference on Circuit System Theory*, Oct. 1965, pp. 621–633.

9. L. B. Jackson, *Digital Filters and Signal Processing*, Kluwer Academic, Norwell, MA, 1996.

10. J. G. Proakis and D. G. Manolakis, *Digital Signal Processing: Principles, Algorithms, and Applications*, Prentice Hall, Upper Saddle River, NJ, 1996.

11. R. G. Lyons, *Understanding Digital Signal Processing*, Addison-Wesley, Reading, MA, 1997.

12. F. J. Harris, On the use of windows for harmonic analysis with the discrete Fourier transform, *Proceedings of the IEEE*, Vol. 66, 1978, pp. 51–83.

13. I. F. Progri, W. R. Michalson, and R. Chassaing, Fast and efficient filter design and implementation on the TMS320C6711 digital signal processor, *2001 International Conference on Acoustics, Speech, and Signal Processing Student Forum*, May 2001.

14. B. Porat, *A Course in Digital Signal Processing*, Wiley, New York, 1997.

15. T. W. Parks and C. S. Burrus, *Digital Filter Design*, Wiley, New York, 1987.

16. S. D. Stearns and R. A. David, *Signal Processing in Fortran and C*, Prentice Hall, Upper Saddle River, NJ, 1993.

17. N. Ahmed and T. Natarajan, *Discrete-Time Signals and Systems*, Reston Publishing, Reston, VA, 1983.

18. S. J. Orfanidis, *Introduction to Signal Processing*, Prentice Hall, Upper Saddle River, NJ, 1996.

19. A. Antoniou, *Digital Filters: Analysis, Design, and Applications*, McGraw-Hill, New York, 1993.

20. E. C. Ifeachor and B. W. Jervis, *Digital Signal Processing: A Practical Approach*, Addison-Wesley, Reading, MA, 1993.

21. P. A. Lynn and W. Fuerst, *Introductory Digital Signal Processing with Computer Applications*, Wiley, New York, 1994.

22. R. D. Strum and D. E. Kirk, *First Principles of Discrete Systems and Digital Signal Processing*, Addison-Wesley, Reading, MA, 1988.

23. D. J. DeFatta, J. G. Lucas, and W. S. Hodgkiss, *Digital Signal Processing: A System Approach*, Wiley, New York, 1988.

24. C. S. Williams, *Designing Digital Filters*, Prentice Hall, Upper Saddle River, NJ, 1986.

25. R. W. Hamming, *Digital Filters*, Prentice Hall, Upper Saddle River, NJ, 1983.

26. S. K. Mitra and J. F. Kaiser, eds., *Handbook for Digital Signal Processing*, Wiley, New York, 1993.

27. S. K. Mitra, *Digital Signal Processing: A Computer-Based Approach*, McGraw-Hill, New York, 1998.

28. R. Chassaing, B. Bitler, and D. W. Horning, Real-time digital filters in C, *Proceedings of the 1991 ASEE Annual Conference*, June 1991.

29. R. Chassaing and P. Martin, Digital filtering with the floating-point TMS320C30 digital signal processor, *Proceedings of the 21st Annual Pittsburgh Conference on Modeling and Simulation*, May 1990.

30. S. D. Stearns and R. A. David, *Signal Processing in Fortran and C*, Prentice Hall, Upper Saddle River, NJ, 1993.

31. R. A. Roberts and C. T. Mullis, *Digital Signal Processing*, Addison-Wesley, Reading, MA, 1987.

32. E. P. Cunningham, *Digital Filtering: An Introduction*, Houghton Mifflin, Boston, 1992.

33. N. J. Loy, *An Engineer's Guide to FIR Digital Filters*, Prentice Hall, Upper Saddle River, NJ, 1988.

34. H. Nuttall, Some windows with very good sidelobe behavior, *IEEE Transactions on Acoustics, Speech, and Signal Processing*, Vol. ASSP-29, No. 1, Feb. 1981.

35. L. C. Ludemen, *Fundamentals of Digital Signal Processing*, Harper & Row, New York, 1986.

36. M. Bellanger, *Digital Processing of Signals: Theory and Practice*, Wiley, New York, 1989.

37. M. G. Bellanger, *Digital Filters and Signal Analysis*, Prentice Hall, Upper Saddle River, NJ, 1986.

38. F. J. Taylor, *Principles of Signals and Systems*, McGraw-Hill, New York, 1994.

39. F. J. Taylor, *Digital Filter Design Handbook*, Marcel Dekker, New York, 1983.

40. W. D. Stanley, G. R. Dougherty, and R. Dougherty, *Digital Signal Processing*, Reston Publishing, Reston, VA, 1984.

41. R. Kuc, *Introduction to Digital Signal Processing*, McGraw-Hill, New York, 1988.

42. H. Baher, *Analog and Digital Signal Processing*, Wiley, New York, 1990.

43. J. R. Johnson, *Introduction to Digital Signal Processing*, Prentice Hall, Upper Saddle River, NJ, 1989.

44. S. Haykin, *Modern Filters*, Macmillan, New York, 1989.

45. T. Young, *Linear Systems and Digital Signal Processing*, Prentice Hall, Upper Saddle River, NJ, 1985.

46. A. Ambardar, *Analog and Digital Signal Processing*, PWS, MA, 1995.

47. A. W. M. van den Enden and N. A. M. Verhoeckx, *Discrete-Time Signal Processing*, Prentice-Hall International, Hemel Hempstead, Hertfordshire, England, 1989.

48. MATLAB, MathWorks, Natick, MA.

49. R. Chassaing and D. W. Horning, *Digital Signal Processing with the TMS320C25*, Wiley, New York, 1990.

5

Infinite Impulse Response Filters

- Infinite impulse response filter structures: direct form I, direct form II, cascade, and parallel
- Bilinear transformation for filter design
- Sinusoidal waveform generation using difference equation
- Filter design and utility packages
- Programming examples using TMS320C6x and C code

The finite impulse response (FIR) filter discussed in Chapter 4 has no analog counterpart. In this chapter we discuss the infinite impulse response (IIR) filter that makes use of the vast knowledge already acquired with analog filters. The design procedure involves the conversion of an analog filter to an equivalent discrete filter using the bilinear transformation (BLT) technique. As such, the BLT procedure converts a transfer function of an analog filter in the s-domain into an equivalent discrete-time transfer function in the z-domain.

5.1 INTRODUCTION

Consider a general input–output equation of the form

$$y(n) = \sum_{k=0}^{N} a_k x(n-k) - \sum_{j=1}^{M} b_j y(n-j) \tag{5.1}$$

$$= a_0 x(n) + a_1 x(n-1) + a_2 x(n-2) + \cdots + a_N x(n-N)$$
$$-b_1 y(n-1) - b_2 y(n-2) - \cdots - b_M y(n-M) \tag{5.2}$$

159

This recursive type of equation represents an infinite impulse response (IIR) filter. The output depends on the inputs as well as past outputs (with feedback). The output $y(n)$, at time n, depends not only on the current input $x(n)$, at time n, and on past inputs $x(n-1)$, $x(n-2)$, ..., $x(n-N)$, but also on past outputs $y(n-1)$, $y(n-2)$, ..., $y(n-M)$.

If we assume all initial conditions to be zero in (5.2), the z-transform of (5.2) becomes

$$Y(z) = a_0 X(z) + a_1 z^{-1} X(z) + a_2 z^{-2} X(z) + \cdots + a_N z^{-N} X(z)$$
$$- b_1 z^{-1} Y(z) - b_2 z^{-2} Y(z) - \cdots - b_M z^{-M} Y(z) \tag{5.3}$$

Let $N = M$ in (5.3); then the transfer function $H(z)$ is

$$H(z) = \frac{Y(z)}{X(z)} = \frac{a_0 + a_1 z^{-1} + a_2 z^{-2} + \cdots + a_N z^{-N}}{1 + b_1 z^{-1} + b_2 z^{-2} + \cdots + b_N z^{-N}} = \frac{N(z)}{D(z)} \tag{5.4}$$

where $N(z)$ and $D(z)$ represent the numerator and denominator polynomial, respectively. Multiplying and dividing by z^N, $H(z)$ becomes

$$H(z) = \frac{a_0 z^N + a_1 z^{N-1} + a_2 z^{N-2} + \cdots + a_N}{z^N + b_1 z^{N-1} + b_2 z^{N-2} + \cdots + b_N} = C \prod_{i=1}^{N} \frac{z - z_i}{z - p_i} \tag{5.5}$$

which is a transfer function with N zeros and N poles. If all the coefficients b_j in (5.5) are zero, this transfer function reduces to the transfer function with N poles at the origin in the z-plane representing the FIR filter discussed in Chapter 4. For a system to be stable, all the poles must reside inside the unit circle, as discussed in Chapter 4. Hence, for an IIR filter to be stable, the magnitude of each of its poles must be less than 1, or:

1. If $|P_i| < 1$, then $h(n) \to 0$, as $n \to \infty$, yielding a stable system.
2. If $|P_i| > 1$, then $h(n) \to \infty$, as $n \to \infty$, yielding an unstable system.

If $|P_i| = 1$, the system is marginally stable, yielding an oscillatory response. Furthermore, multiple-order poles on the unit circle yield an unstable system. Note again that with all the coefficients $b_j = 0$, the system reduces to a nonrecursive and stable FIR filter.

5.2 IIR FILTER STRUCTURES

There are several structures that can represent an IIR filter, as discussed next.

5.2.1 Direct Form I Structure

With the direct form I structure shown in Figure 5.1, the filter in (5.2) can be realized. There is an implied summer (not shown) in Figure 5.1. For an Nth-order filter,

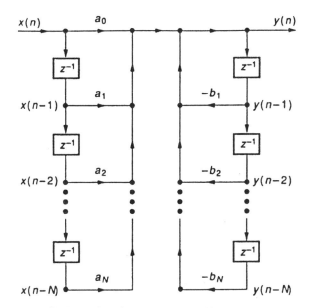

FIGURE 5.1. Direct form I IIR filter structure.

this structure has $2N$ delay elements, represented by z^{-1}. For example, a second-order filter with $N = 2$ will have four delay elements.

5.2.2 Direct Form II Structure

The direct form II structure shown in Figure 5.2 is one of the most commonly used structures. It requires half as many delay elements as the direct form 1. For example, a second-order filter requires two delay elements z^{-1}, as opposed to four with the direct form I. To show that (5.2) can be realized with the direct form II, let a delay variable $U(z)$ be defined as

$$U(z) = \frac{X(z)}{D(z)} \tag{5.6}$$

where $D(z)$ is the denominator polynomial of the transfer function in (5.4). From (5.4) and (5.6), $Y(z)$ becomes

$$Y(z) = \frac{N(z)X(z)}{D(z)} = N(z)U(z)$$
$$= U(z)(a_0 + a_1 z^{-1} + a_2 z^{-2} + \cdots + a_N z^{-N}) \tag{5.7}$$

where $N(z)$ is the numerator polynomial of the transfer function in (5.4). From (5.6)

$$X(z) = U(z)D(z) = U(z)(1 + b_1 z^{-1} + b_2 z^{-2} + \cdots + b_N z^{-N}) \tag{5.8}$$

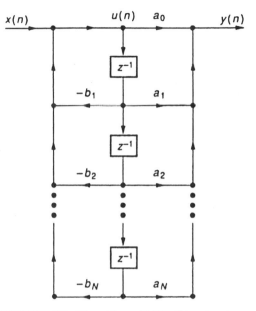

FIGURE 5.2. Direct form II IIR filter structure.

Taking the inverse z-transform of (5.8) yields

$$x(n) = u(n) + b_1 u(n-1) + b_2 u(n-2) + \cdots + b_N u(n-N) \qquad (5.9)$$

Solving for $u(n)$ in (5.9) gives us

$$u(n) = x(n) - b_1 u(n-1) - b_2 u(n-2) - \cdots - b_N u(n-N) \qquad (5.10)$$

Taking the inverse z-transform of (5.7) yields

$$y(n) = a_0 u(n) + a_1 u(n-1) + a_2 u(n-2) + \cdots + a_N u(n-N) \qquad (5.11)$$

The direct form II structure can be represented by (5.10) and (5.11). The delay variable $u(n)$ at the middle top of Figure 5.2 satisfies (5.10), and the output $y(n)$ in Figure 5.2 satisfies (5.11).

Equations (5.10) and (5.11) are used to program an IIR filter. Initially, $u(n-1)$, $u(n-2)$, ... are set to zero. At time n, a new sample $x(n)$ is acquired, and (5.10) is used to solve for $u(n)$. The filter's output at time n then becomes

$$y(n) = a_0 u(n) + 0$$

At time $n+1$, a newer sample $x(n+1)$ is acquired and the delay variables in (5.10) are updated, or

$$u(n+1) = x(n+1) - b_1 u(n) - 0$$

where $u(n-1)$ is updated to $u(n)$. From (5.11), the output at time $n+1$ is

$$y(n+1) = a_0 u(n+1) + a_1 u(n) + 0$$

and so on, for time $n+2, n+3, \ldots$, when for each specific time, a new input sample is acquired and the delay variables and then the output are calculated using (5.10), and (5.11), respectively.

5.2.3 Direct Form II Transpose

The direct form II transpose structure is a modified version of the direct form II and requires the same number of delay elements. The following steps yield a transpose structure from a direct form II version:

1. Reverse the directions of all the branches.
2. Reverse the roles of the input and output (input $\leftrightarrow$ output).
3. Redraw the structure such that the input node is on the left and the output node is on the right (as is typically done).

The direct form II transpose structure is shown in Figure 5.3. To verify this, let $u_0(n)$ and $u_1(n)$ be as shown in Figure 5.3. Then, from the transpose structure,

$$u_0(n) = a_2 x(n) - b_2 y(n) \tag{5.12}$$

$$u_1(n) = a_1 x(n) - b_1 y(n) + u_0(n-1) \tag{5.13}$$

$$y(n) = a_0 x(n) + u_1(n-1) \tag{5.14}$$

Equation (5.13) becomes, using (5.12) to find $u_0(n-1)$,

$$u_1(n) = a_1 x(n) - b_1 y(n) + [a_2 x(n-1) - b_2 y(n-1)] \tag{5.15}$$

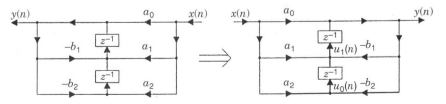

FIGURE 5.3. Direct form II transpose IIR filter structure.

Equation (5.14) becomes, using (5.15) to solve for $u_1(n-1)$,

$$y(n) = a_0 x(n) + [a_1 x(n-1) - b_1 y(n-1) + a_2 x(n-2) - b_2 y(n-2)] \qquad (5.16)$$

which is the same general input–output equation (5.2) for a second-order system. This transposed structure implements first the zeros and then the poles, whereas the direct form II structure implements the poles first.

5.2.4 Cascade Structure

The transfer function in (5.5) can be factored as

$$H(z) = CH_1(z)H_2(z) \cdots H_r(z) \qquad (5.17)$$

in terms of first- or second-order transfer functions. The cascade (or series) structure is shown in Figure 5.4. An overall transfer function can be represented with cascaded transfer functions. For each section, the direct form II structure or its transpose version can be used. Figure 5.5 shows a fourth-order IIR structure in terms of two direct form II second-order sections in cascade. The transfer function $H(z)$, in terms of cascaded second-order transfer functions, can be written as

$$H(z) = \prod_{i=1}^{N/2} \frac{a_{0i} + a_{1i}z^{-1} + a_{2i}z^{-2}}{1 + b_{1i}z^{-1} + b_{2i}z^{-2}} \qquad (5.18)$$

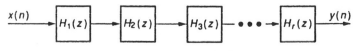

FIGURE 5.4. Cascade form IIR filter structure.

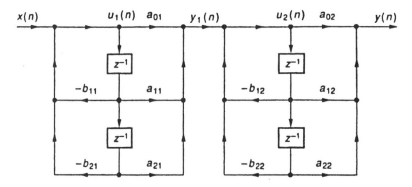

FIGURE 5.5. Fourth-order IIR filter with two direct form II sections in cascade.

where the constant C in (5.17) is incorporated into the coefficients, and each section is represented by i. For example, $N = 4$ for a fourth-order transfer function, and (5.18) becomes

$$H(z) = \frac{(a_{01} + a_{11}z^{-1} + a_{21}z^{-2})(a_{02} + a_{12}z^{-1} + a_{22}z^{-2})}{(1 + b_{11}z^{-1} + b_{21}z^{-2})(1 + b_{12}z^{-1} + b_{22}z^{-2})} \qquad (5.19)$$

as can be verified in Figure 5.5. From a mathematical standpoint, proper ordering of the numerator and denominator factors does not affect the output result. However, from a practical standpoint, proper ordering of each second-order section can minimize quantization noise [1–5]. Note that the output of the first section, $y_1(n)$, becomes the input to the second section. With an intermediate output result stored in one of the registers, a premature truncation of the intermediate output becomes negligible. A programming example will illustrate the implementation of an IIR filter cascaded into second-order direct form II sections.

5.2.5 Parallel Form Structure

The transfer function in (5.5) can be represented as

$$H(z) = C + H_1(z) + H_2(z) + \cdots + H_r(z) \qquad (5.20)$$

which can be obtained using a partial fraction expansion (PFE) on (5.5). This parallel form structure is shown in Figure 5.6. Each of the transfer functions $H_1(z)$,

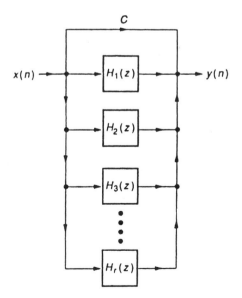

FIGURE 5.6. Parallel form IIR filter structure.

$H_2(z), \ldots$ can be either first- or second-order functions. As with the cascade structure, the parallel form can be efficiently represented in terms of second-order direct form II structure sections. $H(z)$ can be expressed as

$$H(z) = C + \sum_{i=1}^{N/2} \frac{a_{0i} + a_{1i}z^{-1} + a_{2i}z^{-2}}{1 + b_{1i}z^{-1} + b_{2i}z^{-2}} \qquad (5.21)$$

For example, for a fourth-order transfer function, $H(z)$ in (5.21) becomes

$$H(z) = C + \frac{a_{01} + a_{11}z^{-1} + a_{21}z^{-2}}{1 + b_{11}z^{-1} + b_{21}z^{-2}} + \frac{a_{02} + a_{12}z^{-1} + a_{22}z^{-2}}{1 + b_{12}z^{-1} + b_{22}z^{-2}} \qquad (5.22)$$

This fourth-order parallel structure is represented in terms of two direct form II sections as shown in Figure 5.7. From Figure 5.7, the output $y(n)$ can be expressed in terms of the output of each section, or

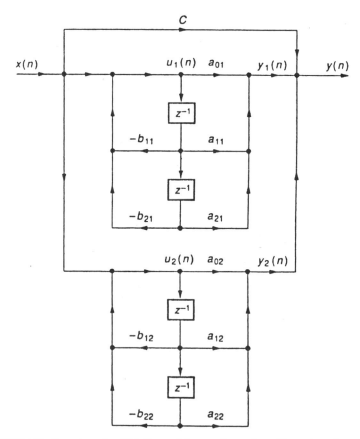

FIGURE 5.7. Fourth-order IIR filter with two direct form II sections in parallel.

$$y(n) = Cx(n) + \sum_{i=1}^{N/2} y_i(n) \tag{5.23}$$

There are other structures, such as the lattice structure, which are useful for applications in speech and adaptive filtering. Although such a structure is not as computationally efficient as the direct form II or cascade structures, requiring more multiplication operations, it is less sensitive to quantization effects [6–8]. The quantization error associated with the coefficients of an IIR filter depends on the amount of shift in the position of its transfer function's poles and zeros in the complex plane. This implies that the shift in the position of a particular pole depends on the position of all the other poles. To minimize this dependency of poles, an Nth-order IIR filter is typically implemented as cascaded second-order sections.

5.3 BILINEAR TRANSFORMATION

The bilinear transformation (BLT) is the most commonly used technique for transforming an analog filter into a discrete filter. It provides a one-to-one mapping from the analog s-plane to the digital z-plane, using

$$s = K \frac{z-1}{z+1} \tag{5.24}$$

The constant K in (5.24) is commonly chosen as $K = 2/T$, where T represents a sampling variable. Other values for K can be selected, since it has no consequence in the design procedure. We choose $T = 2$ or $K = 1$ for convenience, to illustrate the bilinear transformation procedure. Solving for z in (5.24) gives us

$$z = \frac{1+s}{1-s} \tag{5.25}$$

This transformation allows the following:

1. The left region in the s-plane, corresponding to $\sigma < 0$, maps *inside* the unit circle in the z-plane.
2. The right region in the s-plane, corresponding to $\sigma > 0$, maps *outside* the unit circle in the z-plane.
3. The imaginary $j\omega$ axis in the s-plane maps *on* the unit circle in the z-plane.

Let ω_A and ω_D represent the analog and digital frequencies, respectively. With $s = j\omega_A$ and $z = e^{j\omega_D T}$, (5.24) becomes

$$j\omega_A = \frac{e^{j\omega_D T} - 1}{e^{j\omega_D T} + 1} = \frac{e^{j\omega_D T/2}\left(e^{j\omega_D T/2} - e^{-j\omega_D T/2}\right)}{e^{j\omega_D T/2}\left(e^{j\omega_D T/2} + e^{-j\omega_D T/2}\right)} \tag{5.26}$$

Using Euler's expressions for sine and cosine in terms of complex exponential functions, ω_A from (5.26) becomes

$$\omega_A = \tan\frac{\omega_D T}{2} \tag{5.27}$$

which relates the analog frequency ω_A to the digital frequency ω_D. This relationship is plotted in Figure 5.8 for positive values of ω_A. The region corresponding to ω_A between 0 and 1 is mapped into the region corresponding to ω_D between 0 and $\omega_s/4$ in a fairly linear fashion, where ω_s is the sampling frequency in radians. However, the entire region of $\omega_A > 1$ is quite nonlinear, mapping into the region corresponding to ω_D between $\omega_s/4$ and $\omega_s/2$. This compression within this region is referred to as *frequency warping*. As a result, prewarping is done to compensate for this frequency warping. The frequencies ω_A and ω_D are such that

$$H(s)\big|_{s=j\omega_A} = H(z)\big|_{z=e^{j\omega_D T}} \tag{5.28}$$

5.3.1 Bilinear Transformation Design Procedure

The bilinear transformation design procedure makes use of a known analog transfer function for the design of a discrete-time filter. It can be applied using well-documented analog filter functions (Butterworth, Chebychev, etc.). Several types of filter design are available with MATLAB, described in Appendix D. Chebyshev type I and II provide equiripple responses in the passbands and stopbands, respectively. For a given specification, these filters are of lower order than Butterworth-type filters, which have monotonic responses in both passbands and stopbands. An elliptic design has equiripple in both bands and achieves a lower order than a

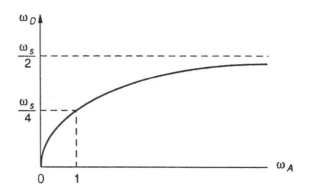

FIGURE 5.8. Relationship between analog and digital frequencies.

Chebyshev-type design; however, it is more difficult to design, with a highly non-linear phase response in the passbands. Although a Butterworth design requires a higher order, it has a linear phase in the passbands.

Perform the following steps in order to use the BLT technique and find $H(z)$.

1. Obtain a known analog transfer function $H(s)$.
2. Prewarp the desired digital frequency ω_D to obtain the analog frequency ω_A in (5.27).
3. Scale the frequency of the analog transfer function $H(s)$ selected, using

$$H(s)|_{s=s/\omega_A} \qquad (5.29)$$

4. Obtain $H(z)$ using the BLT equation (5.24), or

$$H(z) = H(s/\omega_A)|_{s=(z-1)/(z+1)} \qquad (5.30)$$

In the case of bandpass and bandstop filters with lower and upper cutoff frequencies ω_{D1} and ω_{D2}, the two analog frequencies ω_{A1} and ω_{A2} need to be solved. The exercises in Appendix E further illustrate the BLT procedure.

5.4 PROGRAMMING EXAMPLES USING C CODE

Five examples are introduced to illustrate implementation of an IIR filter using the cascaded direct form II structure and the generation of a tone using a difference equation.

Example 5.1: IIR Filter Implementation Using Second-Order Stages in Cascade (IIR)

Figure 5.9 shows a listing of the program IIR.c that implements a generic IIR filter using cascaded second-order stages (sections). The program uses the following two equations associated with each stage:

```
u(n) = x(n) - b₁u(n - 1) - b₂u(n - 2)
y(n) = a₀ u(n) + a₁u(n - 1) + a₂u(n - 2)
```

The loop section of code within the program is processed five times (the number of stages) for each value of n, or sample period. For the first stage, x(n) is the newly acquired input sample. However, for the other stages, the input x(n) is the output y(n) of the preceding stage.

The coefficients b[i][0] and b[i][1] correspond to b_1 and b_2, respectively; where i represents each stage. The delays dly[i][0] and dly[i][1] correspond to u(n - 1) and u(n - 2), respectively.

```
//IIR.c IIR filter using cascaded Direct Form II
//Coefficients a's and b's correspond to b's and a's, from MATLAB

#include "bs1750.cof"                  //BS @ 1750Hz coefficient file
short dly[stages][2] = {0};           //delay samples per stage

interrupt void c_int11()              //ISR
 {
 int i, input;
 int un, yn;

 input = input_sample();              //input to 1st stage
 for (i = 0; i < stages; i++)         //repeat for each stage
  {
    un=input-((b[i][0]*dly[i][0])>>15) - ((b[i][1]*dly[i][1])>>15);

    yn=((a[i][0]*un)>>15)+((a[i][1]*dly[i][0])>>15)+((a[i][2]*dly[i][1])>>15);

    dly[i][1] = dly[i][0];            //update delays
    dly[i][0] = un;                   //update delays
    input = yn;                       //intermediate output->in to next stage
  }
   output_sample(yn);                 //output final result for time n
   return;                            //return from ISR
 }

void main()
{
  comm_intr();                        //init DSK, codec, McBSP
  while(1);                           //infinite loop
}
```

FIGURE 5.9. IIR filter program using second-order sections in cascade (IIR.c).

IIR Bandstop

The coefficient file *bs1750.cof* (Figure 5.10) is obtained from Appendix D. It represents a tenth-order IIR bandstop filter designed with MATLAB's filter designer SPTOOL, as shown in Figure D.2 in Appendix D. Note that MATLAB's filter designer shows the order as 5, which represents the number of second-order stages. The coefficient file contains the numerator coefficients, a's (three per stage), and the denominator coefficients, b's (two per stage). The *a's* and *b's* used in this book correspond to the *b's* and *a's* used in MATLAB.

Build and run this project as **IIR**. Verify that the output is an IIR bandstop filter centered at 1750 Hz. Figure 5.11 shows the output frequency response of this IIR bandstop filter obtained with an HP analyzer (with noise as the input).

IIR Bandpass and Lowpass

1. Rebuild this project using the coefficient file *bp2000.cof* (on the accompanying disk), which represents a 36th-order (18 stages) Chebyshev type 2 IIR

```
//bs1750.cof IIR bandstop coefficient file, centered at 1,750Hz

#define stages 5                    //number of 2nd-order stages

int a[stages][3]=        {          //numerator coefficients
{27940, -10910, 27940},             //a10, a11, a12 for 1st stage
{32768, -11841, 32768},             //a20, a21, a22 for 2nd stage
{32768, -13744, 32768},             //a30, a31, a32 for 3rd stage
{32768, -11338, 32768},             //a40, a41, a42 for 4th stage
{32768, -14239, 32768}  };

int b[stages][2]=        {          //*denominator coefficients
{-11417, 25710},                    //b11, b12 for 1st stage
{-9204, 31581},                     //b21, b22 for 2nd stage
{-15860, 31605},                    //b31, b32 for 3rd stage
{-10221, 32581},                    //b41, b42 for 4th stage
{-15258, 32584}          };         //b51, b52 for 5th stage
```

FIGURE 5.10. Coefficient file for a tenth-order IIR bandstop filter designed with MATLAB in Appendix D (bs1750.cof).

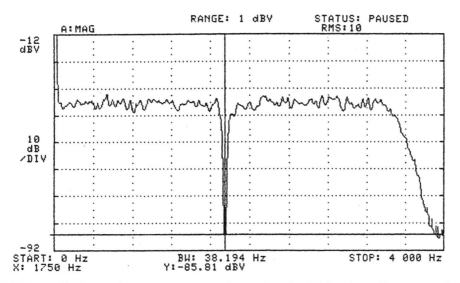

FIGURE 5.11. Output frequency response of a tenth-order IIR bandstop filter centered at 1750 Hz obtained with an HP analyzer.

bandpass filter centered at 2 kHz. This filter was designed with MATLAB, as shown in Figure 5.12. Verify that the filter's output is an IIR bandpass filter centered at 2 kHz. Figure 5.13 shows the output frequency response of this 36th-order IIR bandpass filter, obtained with an HP analyzer.

2. Rebuild this project using the coefficient file *lp2000.cof* (on the disk), which represents an eighth-order IIR lowpass filter with a 2-kHz cutoff

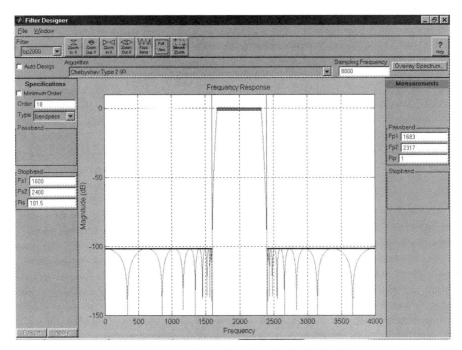

FIGURE 5.12. MATLAB's filter designer (SPTOOL) displaying frequency characteristics of a 36th-order IIR bandpass filter.

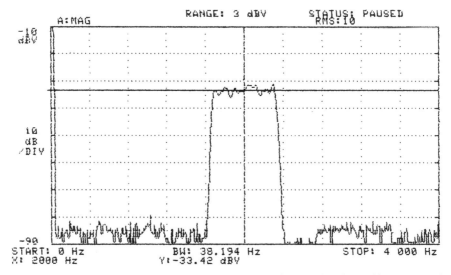

FIGURE 5.13. Output frequency response of a 36th-order IIR bandpass filter centered at 2000 Hz obtained with an HP analyzer.

frequency (also designed with MATLAB). Verify the output of this IIR lowpass filter.

Example 5.2: Generation of Two Tones Using Two Second-Order Difference Equations (two_tones)

This example generates and adds two tones using a difference equation scheme. The output is also stored in memory and plotted within CCS. The difference equation to generate a sine wave is

$$y(n) = Ay(n - 1) - y(n - 2)$$

where

$$A = 2\cos(\omega T)$$
$$y(-1) = -\sin(\omega T)$$
$$y(-2) = -\sin(2\omega T)$$

with two initial conditions, $y(-1)$ and $y(-2)$, $\omega = 2\pi f$, and $T = 1/F_s = 1/(8\,\text{kHz}) = 0.125\,\text{ms}$, the sampling period. The z-transform of $y(n)$ is

$$Y(z) = A\{z^{-1}Y(z) + y(-1)\} - \{z^{-2}Y(z) + z^{-1}y(-1) + y(-2)\}$$

which can be written as

$$\begin{aligned} Y(z)\{1 - Az^{-1} + z^{-2}\} &= Ay(-1) - z^{-1}\,y(-1) - y(-2) \\ &= -2\cos(\omega T)\sin(\omega T) + z^{-1}\sin(\omega T) + \sin(2\omega T) \\ &= z^{-1}\,\sin(\omega T) \end{aligned}$$

Solving for $Y(z)$ yields

$$Y(z) = z\sin(\omega T)/(z^2 - Az + 1)$$

The inverse z-transform of $Y(z)$ is

$$y(n) = ZT^{-1}\{Y(z)\} = \sin(n\omega T)$$

$f = 1.5\,kHz$

$$\begin{aligned} A = 2\cos(\omega T) &= 0.765 \rightarrow A*2^{14} &&= 12{,}540 \\ y(-1) = -\sin(\omega T) &= -0.924 \rightarrow y(-1)*2^{14} &&= -15{,}137 \\ y(-2) = -\sin(2\omega T) &= -0.707 \rightarrow y(-2)*2^{14} &&= -11{,}585 \end{aligned}$$

$f = 2\,kHz$

$$A = 0$$
$$y(-1) = -1 \rightarrow y(-1)*2^{14} = -16384$$
$$y(-2) = 0$$

The coefficient of the second-order difference equation A, along with the two initial conditions, determine the frequency generated. They are scaled for a fixed-point implementation. Using the difference equation

$$y(n) = Ay(n-1) - y(n-2)$$

the output at time $n = 0$ is

$$y(0) = Ay(-1) - y(-2) = -2\cos(\omega T)\sin(\omega T) + \sin(2\omega T) = 0$$

Figure 5.14 shows a listing of the program $two\_tones.c$ that implements a tone generation using a difference equation. The array y1[3] contains the values for $y1(0)$, $y1(-1)$, and $y1(-2)$ to generate a 1.5-kHz tone, and the array y2[3] contains the values for $y2(0)$, $y2(-1)$, and $y2(-2)$ to generate a 2-kHz tone. The function $sinegen$ uses the second-order difference equation to generate each tone, then adds the two tones. Scaling by 2^{14} produces better results for a fixed-point implementation.

Build and run this project as **two_tones**. Verify that the output is the sum of the 1.5- and 2-kHz tones. The output is also stored in a memory buffer. Use CCS to plot the FFT magnitude of the two sinusoids, as shown in Figure 5.15. The starting address of the buffer is sinegen_buffer (see also Example 1.2).

The technique above can be used to generate dual-tone multifrequency: for example, generating and adding the two tones with frequencies of 697 and 1209 Hz, which correspond to the key "3" in a phone.

Example 5.3: Sine Generation Using a Difference Equation (sinegenDE)

This example also generates a sinusoidal tone using an alternative difference equation. See also Example 5.2, which generates/adds two tones. Consider the second-order difference equation obtained in Chapter 4:

$$y(n) = Ay(n-1) + By(n-2) + Cx(n-1)$$

where $B = -1$. Apply an impulse at $n = 1$, so that $x(n-1) = x(0) = 1$, and zero otherwise. For $n = 1$,

$$y(1) = Ay(0) + By(-1) + Cx(0) = C$$

```
//two_tones.c Generates/adds two tones using difference equations

short sinegen(void);                 //for generating tone
short output;                        //for output
short sinegen_buffer[256];           //buffer for output data
const short bufferlength = 256;      //buffer size for plot with CCS
short i = 0;                         //buffer count index

short y1[3] = {0,-15137,-11585};     //y1(0),y1(-1),y1(-2) for 1.5kHz
const short A1 = 12540;              //A1 = 2coswT scaled by 2^14
short y2[3] = {0,-16384,0};          //y2(0),y2(-1),y2(-2) for 2kHz
const short A2 = 0;                  //A2 = 2coswT scaled by 2^14

interrupt void c_int11()             //ISR
{
 output = sinegen();                 //out from tone generation function
 sinegen_buffer[i] = output;         //output into buffer
 output_sample(output);              //output result
 i++;                                //increment buffer count
 if (i == bufferlength) i = 0;       //if buffer count=size of buffer
 return;                             //return to main
}

short sinegen()                                //function to generate tone
{
 y1[0] =((((int)y1[1]*(int)A1))>>14)-y1[2]; //y1(n)=A1*y1(n-1)-y1(n-2)
 y1[2] = y1[1];                                //update y1(n-2)
 y1[1] = y1[0];                                //update y1(n-1)

 y2[0] =((((int)y2[1]*(int)A2))>>14)-y2[2]; //y2(n)=A2*y2(n-1)-y2(n-2)
 y2[2] = y2[1];                                //update y2(n-2)
 y2[1] = y2[0];                                //update y2(n-1)

 return (y1[0] + y2[0]);                       //add the two tones
}

void main()
{
 comm_intr();                        //init DSK, codec, McBSP
 while(1);                           //infinite loop
}
```

FIGURE 5.14. Program to generate and add two tones (two_tones.c).

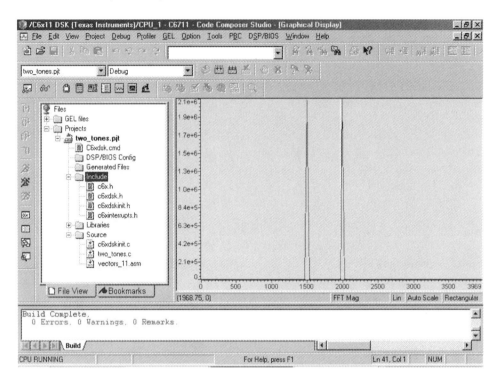

FIGURE 5.15. FFT Magnitude plot of output with two tones using CCS.

with $y(0) = 0$ and $y(-1) = 0$. For $n \geq 2$,

$$y(n) = Ay(n - 1) - y(n - 2).$$

The coefficients $A = 2\cos(\omega T)$ and $C = \sin(\omega T)$ are calculated for a given sampling period $T = 1/F_s$ and a desired frequency ω.

$f = 1.5\,kHz$

$$A = 2\cos(\omega T) = 0.765 \rightarrow A*2^{14} = 12{,}540$$
$$y(1) = C = 0.924 \rightarrow C*2^{14} = 15{,}137$$
$$y(2) = Ay(1) = 0.707 \rightarrow y(2)*2^{14} = 11{,}585$$

$f = 2\,kHz$

$$A = 2\cos(\omega T) = 0$$
$$y(1) = C = \sin(\omega T) = 1 \rightarrow C*2^{14} = 16{,}384$$
$$y(2) = Ay(1) - y(0) = AC = 0$$

Figure 5.16 shows a listing of the program $sinegenDE.c$, which generates a sine wave using this alternative difference equation. This difference equation is calculated within the interrupt service routine (ISR) using an alternative scheme to the

```
//SinegenDE.c  Generates a sinewave using a difference equation

short y[3] = {0,16384,0};                      //y(1) = sinwT
const short A = 0;                             //A = 2*coswT * 2^14
int n = 2;

interrupt void c_int11()                       //ISR
{
 y[n] = (((int)A*(int)y[n-1])>>14) - y[n-2];  //y(n) = Ay(n-1)-y(n-2)
 y[n-2] = y[n-1];                             //update y(n-2)
 y[n-1] = y[n];                               //update y(n-1)
 output_sample(y[n]);                         //output result
 return;                                       //return to main
}

void main()
{
 comm_intr();                                  //init DSK, codec, McBSP
 while(1);                                     //infinite loop
}
```

FIGURE 5.16. Program to generate a sine wave using a difference equation (sine-genDE.c).

implementation in Example 5.2. The coefficient A = 0, and the array y[3], which contains $y(0)$, $y(1)$, and $y(2)$, generate a 2-kHz sine wave.

Build and run this project as **sinegenDE**. Verify that the output is a 2-kHz tone. Change the array to y[3] = {0,15137,11585} and A = 12,540. Rebuild/run the program and verify a 1.5-kHz tone generated at the output. A 3-kHz tone can be generated using $A = -23,170$ and y[3] = {0,11585,0}.

Example 5.4: Generation of a Swept Sinusoid Using a Difference Equation (sweepDE)

Figure 5.17 shows a listing of the program *sweepDE.c*, which generates a sinusoidal signal, sweeping in frequency. The program implements the difference equation

$$y(n) = Ay(n-1) - y(n-2)$$

where $A = 2\cos(\omega T)$ and the two initial conditions are $y(-1) = \sin(\omega T)$ and $y(-2) = -\sin(2\omega T)$. Example 5.2 illustrates the generation of a sine wave using this difference equation.

An initial signal frequency is set in the program at 500 Hz. The signal's frequency is incremented by 10 Hz until a set maximum frequency of 3500 Hz is reached. The

```
//SweepDE.c Generates a sweeping sinusoid using a difference equation

#include <math.h>
#define   two_pi   (2*3.1415926)      //2*pi
#define   two_14            16384      //2^14
#define   T             0.000125      //sample period = 1/Fs
#define   MIN_FREQ           500      //initial frequency of sweep
#define   MAX_FREQ          3500      //max frequency of sweep
#define   STEP_FREQ           10      //step frequency
#define   SWEEP_PERIOD       200      //lasting time at one frequency
short     y0 = 0;                     //initial output
short     y_1 = -6270;                //y(-1)=-sinwT(scaled) f=500Hz
short     y_2 = -11585;               //y(-2_=-sin2wT(scaled) f=500Hz
short     A = 30274;                  //A = 2*coswT scaled by 2^14
short     freq = MIN_FREQ;            //current frequency
short     sweep_count = 0;            //counter for lasting time
void      coeff_gen(short);           //function prototype for coeff

interrupt void c_int11()              //ISR
{
 sweep_count++;                       //incr lasting time at one freq
 if(sweep_count >= SWEEP_PERIOD)      //time reaches max duration
   {
    if(freq >= MAX_FREQ)              //if the current frequency is max
       freq = MIN_FREQ;              //reinit to initial frequency
    else
       freq = freq + STEP_FREQ;      //incr to next higher frequency

    coeff_gen(freq);                 //function for new set of coeff
    sweep_count = 0;                 //reset counter for lasting time
   }
 y0=(((int)A * (int)y_1)>>14) - y_2; //y(n) = A*y(n-1) - y(n-2)
 y_2 = y_1;                           //update y(n-2)
 y_1 = y0;                            //update y(n-1)
 output_sample(y0);                   //output result
}

void coeff_gen(short freq)            //calculate new set of coeff
{
 float w;                             //angular frequency

 w = two_pi*freq;                     //w = 2*pi*f
 A = 2*cos(w*T)*two_14;               //A = 2*coswT * (2^14)
 y_1 = -sin(w*T)*two_14;              //y_1 = -sinwT *(2^14)
 y_2 = -sin(2*T*w)*two_14;            //y_2 = -sin2wT * (2^14)
 return;
}

void main()
{
 comm_intr();                         //init DSK, codec, McBSP
 while(1);                            //infinite loop
}
```

FIGURE 5.17. Program to generate a sweeping sinusoid using a difference equation
(sweepDE.c).

178

duration of the sinusoidal signal at each frequency generated is set with 200 and can be reduced for a faster sweep.

With an initial frequency of 500 Hz, the constants A $= 30,274$, $y(0) = 0$, $y(-1) = -6270$ and $y(-2) = -11,585$ (see Example 5.2). For each frequency (510, 520, . . .) the function coeff_gen is called to calculate a new set of constants A, $y(n - 1)$, $y(n - 2)$ to implement the difference equation. A slider can be used to control the swept signal, such as the step or incremental frequency and the duration of the sinusoidal signal at each incremental frequency.

Build and run this project as **sweepDE**. Verify that the output is a swept sinusoidal signal.

Example 5.5: IIR Inverse Filter (IIRinverse)

This example illustrates an IIR inverse filter. With noise as input, a forward IIR filter is calculated. The output of the forward filter becomes the input to an inverse IIR filter. The output of the inverse filter is the original input noise sequence. See Example 4.10, which implements an inverse FIR filter, and Example 5.1, which implements an IIR filter.

The transfer function of an IIR filter is

$$H(z) = \frac{\sum_{i=0}^{N-1} a_i z^{-i}}{\sum_{j=1}^{M-1} b_j z^{-j}}$$

The output sequence of the IIR filter is

$$y(n) = \sum_{i=0}^{N-1} a_i x(n-i) - \sum_{j=1}^{M-1} b_j y(n-j)$$

where $x(n - i)$ represents the input sequence. The input sequence $x(n)$ can then be recovered using $\hat{x}(n)$ as an estimate of $x(n)$, or

$$\hat{x}(n) = \frac{y(n) + \sum_{j=1}^{M-1} b_j y(n-j) - \sum_{i=1}^{N-1} a_i \hat{x}(n-i)}{a_0}$$

The program **IIRinverse.c** (Figure 5.18) implements the inverse IIR filter.

Build this project as **IIRinverse**. Use noise as input to the system (from Goldwave, noise generator, etc.). Run the program and verify that the resulting output is the input noise (with the slider in the default position 1).

//IIRinverse.C Inverse IIR Filter

```
#include "bp2000.cof"              //BP @ 2 kHz coefficient file
short dly[stages][2] = {0};       //delay samples per stage
short out_type = 1;               //type of output for slider
short a0, a1, a2, b1, b2;         //coefficients

interrupt void c_int11()          //ISR
{
 short i, input, input1;
 int un1, yn1, un2, input2, yn2;

 input1 = input_sample();         //input to 1st stage
 input = input1;                  //original input
 for(i = 0; i < stages; i++)      //repeat for each stage
  {
   a1 = ((a[i][1]*dly[i][0])>>15); //a1*u(n-1)
   a2 = ((a[i][2]*dly[i][1])>>15); //a2*u(n-2)
   b1 = ((b[i][0]*dly[i][0])>>15); //b1*u(n-1)
   b2 = ((b[i][1]*dly[i][1])>>15); //b2*u(n-2)
   un1 = input1 - b1 - b2;
   a0=((a[i][0]*un1)>>15);

   yn1 = a0 + a1 + a2;            //stage output
   input1 = yn1;                  //intermediate out->in next stage
   dly[i][1] = dly[i][0];         //update delays u(n-2) = u(n-1)
   dly[i][0] = un1;               //update delays u(n-1) = u(n)
  }
 input2 = yn1;                    //out forward=in reverse filter

 for(i = stages; i > 0; i--)      //for inverse IIR filter
  {
   a1 = ((a[i][1]*dly[i][0])>>15); //a1u(n-1)
   a2 = ((a[i][2]*dly[i][1])>>15); //a2u(n-2)
   b1 = ((b[i][0]*dly[i][0])>>15); //b1u(n-1)
   b2 = ((b[i][1]*dly[i][1])>>15); //b2u(n-2)
   un2 = input2 - a1 - a2;
   yn2 = (un2 + b1 + b2);
   input2 = (yn2<<15)/a[i][0];    //intermediate out->in next stage
  }
 if(out_type == 1)               //if slider in position 1
    output_sample(input);        //original input signal
 if(out_type == 2)
    output_sample(yn1);          //output of forward filter
 if(out_type == 3)
    output_sample(yn2 >>6);      //output of inverse filter
 return;                         //return from ISR
}

void main()
{
 comm_intr();                    //init DSK, codec, McBSP
 while(1);                       //infinite loop
}
```

FIGURE 5.18. Program to implement an inverse IIR filter (IIRinverse.c).

Change the slider to position 2 and verify that the output of the forward IIR filter is an IIR bandpass filter centered at 2 kHz. The coefficient file `bp2000.cof` was used in Example 5.1 to implement an IIR filter. With the slider in position 3, verify that the output of the inverse IIR filter is the original input noise.

In this example, the forward filter's characteristics are known. This example can be extended so that the filter's characteristics are unknown. In such a case, the unknown forward filter's coefficients, a's and b's, can be estimated using Prony's method [9].

REFERENCES

1. L. B. Jackson, *Digital Filters and Signal Processing*, Kluwer Academic, Norwell, MA, 1996.

2. L. B. Jackson, Roundoff noise analysis for fixed-point digital filters realized in cascade or parallel form, *IEEE Transactions on Audio and Electroacoustics*, Vol. AU-18, June 1970, pp. 107–122.

3. L. B. Jackson, An analysis of limit cycles due to multiplicative rounding in recursive digital filters, *Proceedings of the 7th Allerton Conference on Circuit and System Theory*, 1969, pp. 69–78.

4. L. B. Lawrence and K. V. Mirna, A new and interesting class of limit cycles in recursive digital filters, *Proceedings of the IEEE International Symposium on Circuit and Systems*, Apr. 1977, pp. 191–194.

5. R. Chassaing and D. W. Horning, *Digital Signal Processing with the TMS320C25*, Wiley, New York, 1990.

6. A. H. Gray and J. D. Markel, Digital lattice and ladder filter synthesis, *IEEE Transactions on Acoustics, Speech, and Signal Processing*, Vol. ASSP-21, 1973, pp. 491–500.

7. A. H. Gray and J. D. Markel, A normalized digital filter structure, *IEEE Transactions on Acoustics, Speech, and Signal Processing*, Vol. ASSP-23, 1975, pp. 268–277.

8. A. V. Oppenheim and R. Schafer, *Discrete-Time Signal Processing*, Prentice Hall, Upper Saddle River, NJ, 1989.

9. I. Progri, W. R. Michalson, and R. Chassaing, Fast and efficient filter design and implementation on the TMS320C6711 digital signal processor, *International Conference on Acoustics Speech and Signal Processing (ICASSP)*, 2001.

10. N. Ahmed and T. Natarajan, *Discrete-Time Signals and Systems*, Reston Publishing, Reston, VA, 1983.

11. D. W. Horning and R. Chassaing, IIR filter scaling for real-time digital signal processing, *IEEE Transactions on Education*, Feb. 1991.

6

Fast Fourier Transform

- The fast Fourier transform using radix-2 and radix-4
- Decimation or decomposition in frequency and in time
- Programming examples

The fast Fourier transform (FFT) is an efficient algorithm that is used for converting a time-domain signal into an equivalent frequency-domain signal, based on the discrete Fourier transform (DFT). Several real-time programming examples on FFT are included.

6.1 INTRODUCTION

The discrete Fourier transform converts a time-domain sequence into an equivalent frequency-domain sequence. The inverse discrete Fourier transform performs the reverse operation and converts a frequency-domain sequence into an equivalent time-domain sequence. The fast Fourier transform (FFT) is a very efficient algorithm technique based on the discrete Fourier transform but with fewer computations required. The FFT is one of the most commonly used operations in digital signal processing to provide a frequency spectrum analysis [1–6]. Two different procedures are introduced to compute an FFT: the decimation-in-frequency and the decimation-in-time. Several variants of the FFT have been used, such as the Winograd transform [7,8], the discrete cosine transform (DCT) [9], and the discrete Hartley transform [10–12]. Programs based on the DCT, FHT, and the FFT are available in Ref. 9.

6.2 DEVELOPMENT OF THE FFT ALGORITHM WITH RADIX-2

The FFT reduces considerably the computational requirements of the discrete Fourier transform (DFT). The DFT of a discrete-time signal $x(nT)$ is

$$X(k) = \sum_{n=0}^{N-1} x(n)W^{nk} \qquad k = 0, 1, \ldots, N-1 \tag{6.1}$$

where the sampling period T is implied in $x(n)$ and N is the frame length. The constants W are referred to as *twiddle constants* or *factors*, which represent the phase, or

$$W = e^{-j2\pi/N} \tag{6.2}$$

and is a function of the length N. Equation (6.1) can be written for $k = 0, 1, \ldots, N-1$, as

$$X(k) = x(0) + x(1)W^k + x(2)W^{2k} + \cdots + x(N-1)W^{(N-1)k} \tag{6.3}$$

This represents a matrix of $N \times N$ terms, since $X(k)$ needs to be calculated for N values for k. Since (6.3) is an equation in terms of a complex exponential, for each specific k there are $(N-1)$ complex additions and N complex multiplications. This results in a total of $(N^2 - N)$ complex additions and N^2 complex multiplications. Hence, the computational requirements of the DFT can be very intensive, especially for large values of N. FFT reduces computational complexity from N^2 to $N \log N$.

The FFT algorithm takes advantage of the periodicity and symmetry of the twiddle constants to reduce the computational requirements of the FFT. From the periodicity of W,

$$W^{k+N} = W^k \tag{6.4}$$

and from the symmetry of W,

$$W^{k+N/2} = -W^k \tag{6.5}$$

Figure 6.1 illustrates the properties of the twiddle constants W for $N = 8$. For example, let $k = 2$, and note that from (6.4), $W^{10} = W^2$, and from (6.5), $W^6 = -W^2$.

For a radix-2 (base 2), the FFT decomposes an N-point DFT into two $(N/2)$-point or smaller DFTs. Each $(N/2)$-point DFT is further decomposed into two $(N/4)$-point DFTs, and so on. The last decomposition consists of $(N/2)$ two-point DFTs. The smallest transform is determined by the radix of the FFT. For a radix-2 FFT, N must be a power or base of 2, and the smallest transform or the last decomposition is the two-point DFT. For a radix-4, the last decomposition is a four-point DFT.

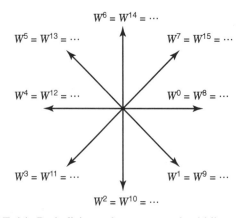

FIGURE 6.1. Periodicity and symmetry of twiddle constant W.

6.3 DECIMATION-IN-FREQUENCY FFT ALGORITHM WITH RADIX-2

Let a time-domain input sequence $x(n)$ be separated into two halves:

(a)
$$x(0), x(1), \ldots, x\left(\frac{N}{2}-1\right) \tag{6.6}$$

and

(b)
$$\left(\frac{N}{2}\right), x\left(\frac{N}{2}+1\right), \ldots, x(N-1) \tag{6.7}$$

Taking the DFT of each set of the sequence in (6.6) and (6.7) gives us

$$X(k) = \sum_{n=0}^{(N/2)-1} x(n)W^{nk} + \sum_{n=N/2}^{N-1} x(n)W^{nk} \tag{6.8}$$

Let $n = n + N/2$ in the second summation of (6.8); $X(k)$ becomes

$$X(k) = \sum_{n=0}^{(N/2)-1} x(n)W^{nk} + W^{kN/2} \sum_{n=0}^{(N/2)-1} x\left(n+\frac{N}{2}\right)W^{nk} \tag{6.9}$$

where $W^{kN/2}$ is taken out of the second summation because it is not a function of n. Using

$$W^{kN/2} = e^{-jk\pi} = \left(e^{-j\pi}\right)^k = \left(\cos\pi - j\sin\pi\right)^k = (-1)^k$$

in (6.9), $X(k)$ becomes

$$X(k) = \sum_{n=0}^{(N/2)-1}\left[x(n)+(-1)^k x\left(n+\frac{N}{2}\right)\right]W^{nk} \tag{6.10}$$

Because $(-1)^k = 1$ for even k and -1 for odd k, (6.10) can be separated for even and odd k, or

1. For even k:

$$X(k) = \sum_{n=0}^{(N/2)-1}\left[x(n)+x\left(n+\frac{N}{2}\right)\right]W^{nk} \tag{6.11}$$

2. For odd k:

$$X(k) = \sum_{n=0}^{(N/2)-1}\left[x(n)-x\left(n+\frac{N}{2}\right)\right]W^{nk} \tag{6.12}$$

Substituting $k = 2k$ for even k, and $k = 2k + 1$ for odd k, (6.11) and (6.12) can be written for $k = 0, 1, \ldots, (N/2) - 1$ as

$$X(2k) = \sum_{n=0}^{(N/2)-1}\left[x(n)+x\left(n+\frac{N}{2}\right)\right]W^{2nk} \tag{6.13}$$

$$x(2k+1) = \sum_{n=0}^{(N/2)-1}\left[x(n)-x\left(n+\frac{N}{2}\right)\right]W^n W^{2nk} \tag{6.14}$$

Because the twiddle constant W is a function of the length N, it can be represented as W_N. Then W_N^2 can be written as $W_{N/2}$. Let

$$a(n) = x(n)+x(n+N/2) \tag{6.15}$$

$$b(n) = x(n)-x(n+N/2) \tag{6.16}$$

Equations (6.13) and (6.14) can be written more clearly as two $(N/2)$-point DFTs, or

$$X(2k) = \sum_{n=0}^{(N/2)-1} a(n)W_{N/2}^{nk} \tag{6.17}$$

$$X(2k+1) = \sum_{n=0}^{(N/2)-1} b(n)W_N^n W_{N/2}^{nk} \tag{6.18}$$

Figure 6.2 shows the decomposition of an N-point DFT into two $(N/2)$-point DFTs, for $N = 8$. As a result of the decomposition process, the X's in Figure 6.2 are even

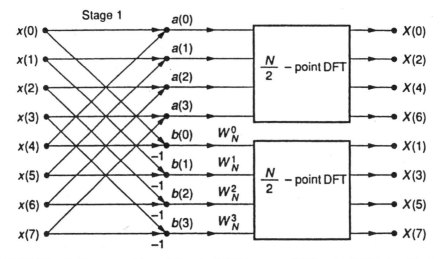

FIGURE 6.2. Decomposition of N-point DFT into two $(N/2)$-point DFTs, for $N = 8$.

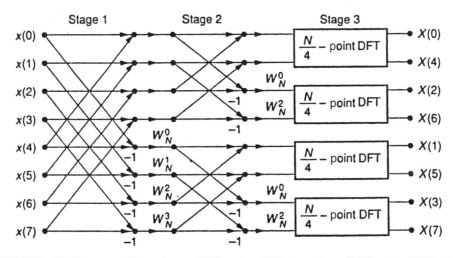

FIGURE 6.3. Decomposition of two $(N/2)$-point DFTs into four $(N/4)$-point DFTs, for $N = 8$.

in the upper half and odd in the lower half. The decomposition process can now be repeated such that each of the $(N/2)$-point DFTs is further decomposed into two $(N/4)$-point DFTs, as shown in Figure 6.3, again using $N = 8$ to illustrate.

The upper section of the output sequence in Figure 6.2 yields the sequence $X(0)$ and $X(4)$ in Figure 6.3, ordered as even. $X(2)$ and $X(6)$ from Figure 6.3 represent the odd values. Similarly, the lower section of the output sequence in Figure 6.2 yields $X(1)$ and $X(5)$, ordered as the even values, and $X(3)$ and $X(7)$ as the odd values. This scrambling is due to the decomposition process. The final order of the

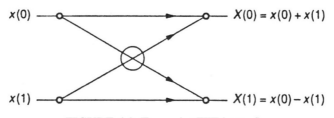

FIGURE 6.4. Two-point FFT butterfly.

output sequence $X(0)$, $X(4)$, ... in Figure 6.3 is shown to be scrambled. The output needs to be resequenced or reordered. Programming examples presented later in this chapter include the appropriate function for resequencing. The output sequence $X(k)$ represents the DFT of the time sequence $x(n)$.

This is the last decomposition, since we now have a set of $(N/2)$ two-point DFTs, the lowest decomposition for a radix-2. For the two-point DFT, $X(k)$ in (6.1) can be written as

$$X(k) = \sum_{n=0}^{1} x(n)W^{nk} \qquad k = 0, 1 \qquad (6.19)$$

or

$$X(0) = x(0)W^0 + x(1)W^0 = x(0) + x(1) \qquad (6.20)$$

$$X(1) = x(0)W^0 + x(1)W^0 = x(0) - x(1) \qquad (6.21)$$

since $W^1 = e^{-j2\pi/2} = -1$. Equations (6.20) and (6.21) can be represented by the flow graph in Figure 6.4, usually referred to as a *butterfly*. The final flow graph of an eight-point FFT algorithm is shown in Figure 6.5. This algorithm is referred as *decimation-in-frequency* (DIF) because the output sequence $X(k)$ is decomposed (decimated) into smaller subsequences, and this process continues through M stages or iterations, where $N = 2^M$. The output $X(k)$ is complex with both real and imaginary components, and the FFT algorithm can accommodate either complex or real input values.

The FFT is not an approximation of the DFT. It yields the same result as the DFT with fewer computations required. This reduction becomes more and more important with higher-order FFT.

There are other FFT structures that have been used to illustrate the FFT. An alternative flow graph to that in Figure 6.5 can be obtained with ordered output and scrambled input.

An eight-point FFT is illustrated through the following exercise. We will see that flow graphs for higher-order FFT (larger N) can readily be obtained.

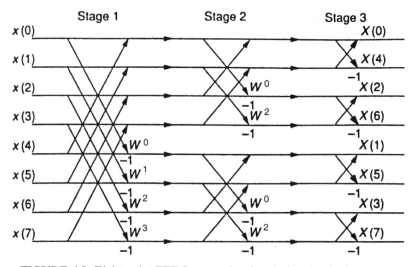

FIGURE 6.5. Eight-point FFT flow graph using decimation-in-frequency.

Exercise 6.1: Eight-Point FFT Using Decimation-in-Frequency

Let the input $x(n)$ represent a rectangular waveform, or $x(0) = x(1) = x(2) = x(3)$ = 1 and $x(4) = x(5) = x(6) = x(7) = 0$. The eight-point FFT flow graph in Figure 6.5 can be used to find the output sequence $X(k), k = 0, 1, \ldots, 7$. With $N = 8$, four twiddle constants need to be calculated, or

$$W^0 = 1$$
$$W^1 = e^{-j2\pi/8} = \cos(\pi/4) - j\sin(\pi/4) = 0.707 - j0.707$$
$$W^2 = e^{-j4\pi/8} = -j$$
$$W^3 = e^{-j6\pi/8} = -0.707 - j0.707$$

The intermediate output sequence can be found after each stage.

Stage 1

$$x(0) + x(4) = 1 \rightarrow x'(0)$$
$$x(1) + x(5) = 1 \rightarrow x'(1)$$
$$x(2) + x(6) = 1 \rightarrow x'(2)$$
$$x(3) + x(7) = 1 \rightarrow x'(3)$$
$$[x(0) - x(4)]W^0 = 1 \rightarrow x'(4)$$
$$[x(1) - x(5)]W^1 = 0.707 - j0.707 \rightarrow x'(5)$$
$$[x(2) - x(6)]W^2 = -j \rightarrow x'(6)$$
$$[x(3) - x(7)]W^3 = -0.707 - j0.707 \rightarrow x'(7)$$

where $x'(0), x'(1), \ldots, x'(7)$ represent the intermediate output sequence after the first iteration, which becomes the input to the second stage.

Stage 2

$$x'(0) + x'(2) = 2 \rightarrow x''(0)$$
$$x'(1) + x'(3) = 2 \rightarrow x''(1)$$
$$[x'(0) - x'(2)]W^0 = 0 \rightarrow x''(2)$$
$$[x'(1) - x'(3)]W^2 = 0 \rightarrow x''(3)$$
$$x'(4) + x'(6) = 1 - j \rightarrow x''(4)$$
$$x'(5) + x'(7) = (0.707 - j0.707) + (-0.707 - j0.707) = -j1.41 \rightarrow x''(5)$$
$$[x'(4) - x'(6)]W^0 = 1 + j \rightarrow x''(6)$$
$$[x'(5) - x'(7)]W^2 = -j1.41 \rightarrow x''(7)$$

The resulting intermediate, second-stage output sequence $x''(0), x''(1), \ldots, x''(7)$ becomes the input sequence to the third stage.

Stage 3

$$X(0) = x''(0) + x''(1) = 4$$
$$X(4) = x''(0) - x''(1) = 0$$
$$X(2) = x''(2) + x''(3) = 0$$
$$X(6) = x''(2) - x''(3) = 0$$
$$X(1) = x''(4) + x''(5) = (1 - j) + (-j1.41) = 1 - j2.41$$
$$X(5) = x''(4) - x''(5) = 1 + j0.41$$
$$X(3) = x''(6) + x''(7) = (1 + j) + (-j1.41) = 1 - j0.41$$
$$X(7) = x''(6) - x''(7) = 1 + j2.41$$

We now use the notation of X's to represent the final output sequence. The values $X(0), X(1), \ldots, X(7)$ form the scrambled output sequence. These results can be verified with MATLAB, described in Appendix D. We show later how to reorder the output sequence and plot the output magnitude.

Exercise 6.2: Sixteen-Point FFT

Given $x(0) = x(1) = \ldots = x(7) = 1$, and $x(8) = x(9) = \ldots x(15) = 0$, which represents a rectangular input sequence. The output sequence can be found using the 16-point flow graph shown in Figure 6.6. The intermediate output results after each stage are found in a manner similar to that in Exercise 6.1. Eight twiddle constants W^0, $W^1, \ldots, W^7$ need to be calculated for $N = 16$.

Verify the scrambled output sequence X's as shown in Figure 6.6. Reorder this output sequence and take its magnitude. Verify the plot in Figure 6.7, which

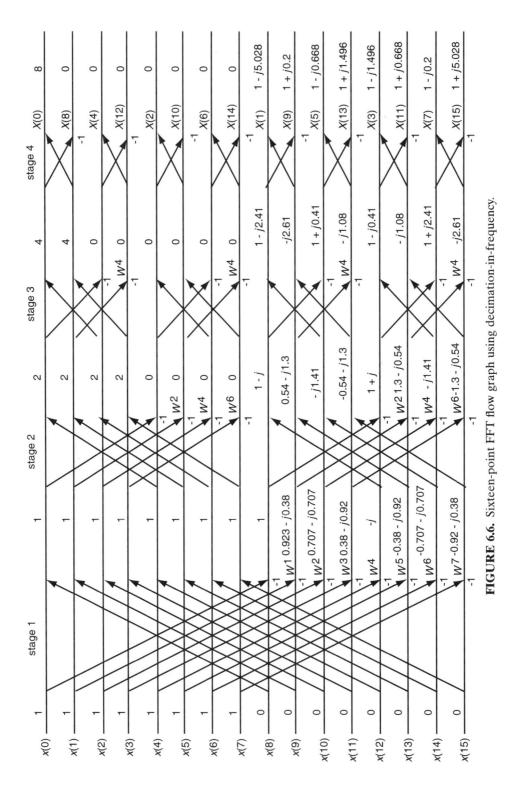

FIGURE 6.6. Sixteen-point FFT flow graph using decimation-in-frequency.

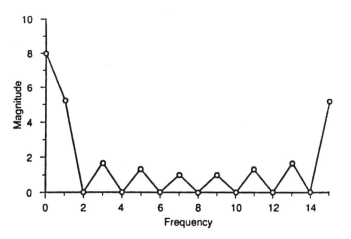

FIGURE 6.7. Output magnitude for 16-point FFT.

represents a sinc function. The output $X(8)$ represents the magnitude at the Nyquist frequency. These results can be verified with MATLAB, described in Appendix D.

6.4 DECIMATION-IN-TIME FFT ALGORITHM WITH RADIX-2

Whereas the decimation-in-frequency (DIF) process decomposes an output sequence into smaller subsequences, the *decimation-in-time* (DIT) is a process that decomposes the input sequence into smaller subsequences. Let the input sequence be decomposed into an even sequence and an odd sequence, or

$$x(0), x(2), x(4), \ldots, x(2n)$$

and

$$x(1), x(3), x(5), \ldots, x(2n+1)$$

We can apply (6.1) to these two sequences to obtain

$$X(k) = \sum_{n=0}^{(N/2)-1} x(2n)W^{2nk} + \sum_{n=0}^{(N/2)-1} x(2n+1)W^{(2n+1)k} \tag{6.22}$$

Using $W_N^2 = W_{N/2}$ in (6.22) yields

$$X(k) = \sum_{n=0}^{(N/2)-1} x(2n)W_{N/2}^{nk} + W_N^k \sum_{n=0}^{(N/2)-1} x(2n+1)W_{N/2}^{nk} \tag{6.23}$$

which represents two $(N/2)$-point DFTs. Let

$$C(k) = \sum_{n=0}^{(N/2)-1} x(2n)W_{N/2}^{nk} \qquad (6.24)$$

$$D(k) = \sum_{n=0}^{(N/2)-1} X(2n+1)W_{N/2}^{nk} \qquad (6.25)$$

Then $X(k)$ in (6.23) can be written as

$$X(k) = C(k) + W_N^k D(k) \qquad (6.26)$$

Equation (6.26) needs to be interpreted for $k > (N/2) - 1$. Using the symmetry property (6.5) of the twiddle constant, $W^{k+N/2} = -W^k$,

$$X(k + N/2) = C(k) - W^k D(k) \qquad k = 0, 1, \ldots, (N/2) - 1 \qquad (6.27)$$

For example, for $N = 8$, (6.26) and (6.27) become

$$X(k) = C(k) + W^k D(k) \qquad k = 0, 1, 2, 3 \qquad (6.28)$$

$$X(k + 4) = C(k) - W^k D(k) \qquad k = 0, 1, 2, 3 \qquad (6.29)$$

Figure 6.8 shows the decomposition of an eight-point DFT into two four-point DFTs with the decimation-in-time procedure. This decomposition or decimation process is repeated so that each four-point DFT is further decomposed into two two-point DFTs, as shown in Figure 6.9. Since the last decomposition is $(N/2)$ two-point DFTs, this is as far as this process goes.

Figure 6.10 shows the final flow graph for an eight-point FFT using a decimation-in-time process. The input sequence is shown to be scrambled in Figure 6.10, in the same manner as the output sequence $X(k)$ was scrambled during the decimation-

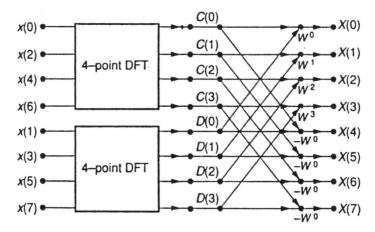

FIGURE 6.8. Decomposition of eight-point DFT into four-point DFTs using DIT.

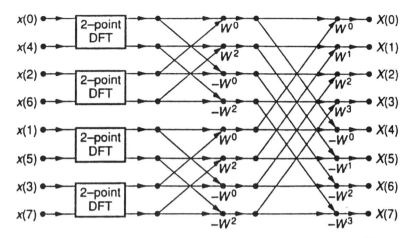

FIGURE 6.9. Decomposition of two four-point DFTs into four two-point DFTs using DIT.

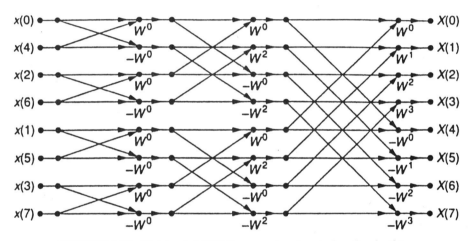

FIGURE 6.10. Eight-point FFT flow graph using decimation-in-time.

in-frequency process. With the input sequence $x(n)$ scrambled, the resulting output sequence $X(k)$ becomes properly ordered. Identical results are obtained with an FFT using either the decimation-in-frequency (DIF) or the decimation-in-time (DIT) process. An alternative DIT flow graph to the one shown in Figure 6.10, with ordered input and scrambled output, can also be obtained.

The following exercise shows that the same results are obtained for an eight-point FFT with the DIT process as in Exercise 6.1 with the DIF process.

Exercise 6.3: Eight-Point FFT Using Decimation-in-Time

Given the input sequence $x(n)$ representing a rectangular waveform as in Exercise 6.1, the output sequence $X(k)$, using the DIT flow graph in Figure 6.10, is the same as in Exercise 6.1. The twiddle constants are the same as in Exercise 6.1. Note that the twiddle constant W is multiplied with the second term only (not with the first).

Stage 1

$$x(0) + W^0 x(4) = 1 + 0 = 1 \rightarrow x'(0)$$
$$x(0) - W^0 x(4) = 1 - 0 = 1 \rightarrow x'(4)$$
$$x(2) + W^0 x(6) = 1 + 0 = 1 \rightarrow x'(2)$$
$$x(2) - W^0 x(6) = 1 - 0 = 1 \rightarrow x'(6)$$
$$x(1) + W^0 x(5) = 1 + 0 = 1 \rightarrow x'(1)$$
$$x(1) - W^0 x(5) = 1 - 0 = 1 \rightarrow x'(5)$$
$$x(3) + W^0 x(7) = 1 + 0 = 1 \rightarrow x'(3)$$
$$x(3) - W^0 x(7) = 1 - 0 = 1 \rightarrow x'(7)$$

where the sequence x's represents the intermediate output after the first iteration and becomes the input to the subsequent stage.

Stage 2

$$x'(0) + W^0 x'(2) = 1 + 1 = 2 \rightarrow x''(0)$$
$$x'(4) + W^2 x'(6) = 1 + (-j) = 1 - j \rightarrow x''(4)$$
$$x'(0) - W^0 x'(2) = 1 - 1 = 0 \rightarrow x''(2)$$
$$x'(4) - W^2 x'(6) = 1 - (-j) = 1 + j \rightarrow x''(6)$$
$$x'(1) + W^0 x'(3) = 1 + 1 = 2 \rightarrow x''(1)$$
$$x'(5) + W^2 x'(7) = 1 + (-j)(1) = 1 - j \rightarrow x''(5)$$
$$x'(1) - W^0 x'(3) = 1 - 1 = 0 \rightarrow x''(3)$$
$$x'(5) - W^2 x'(7) = 1 - (-j)(1) = 1 + j \rightarrow x''(7)$$

where the intermediate second-stage output sequence x''s becomes the input sequence to the final stage.

Stage 3

$$X(0) = x''(0) + W^0 x''(1) = 4$$
$$X(1) = x''(4) + W^1 x''(5) = 1 - j2.414$$
$$X(2) = x''(2) + W^2 x''(3) = 0$$
$$X(3) = x''(6) + W^3 x''(7) = 1 - j0.414$$
$$X(4) = x''(0) - W^0 x''(1) = 0$$
$$X(5) = x''(4) - W^1 x''(5) = 1 + j0.414$$
$$X(6) = x''(2) - W^2 x''(3) = 0$$
$$X(7) = x''(6) - W^3 x''(7) = 1 + j2.414$$

which is the same output sequence as found in Exercise 6.1.

6.5 BIT REVERSAL FOR UNSCRAMBLING

A bit-reversal procedure allows a scrambled sequence to be reordered. To illustrate this bit-swapping process, let $N = 8$, represented by three bits. The first and third bits are swapped. For example, $(100)_b$ is replaced by $(001)_b$. As such, $(100)_b$ specifying the address of $X(4)$ is replaced by or swapped with $(001)_b$ specifying the address of $X(1)$. Similarly, $(110)_b$ is replaced/swapped with $(011)_b$, or the addresses of $X(6)$ and $X(3)$ are swapped. In this fashion, the output sequence in Figure 6.5 with the DIF, or the input sequence in Figure 6.10 with the DIT, can be reordered.

This bit-reversal procedure can be applied for larger values of N. For example, for $N = 64$, represented by six bits, the first and sixth bits, the second and fifth bits, and the third and fourth bits are swapped.

Several examples in this chapter illustrate the FFT algorithm, incorporating algorithms for unscrambling.

6.6 DEVELOPMENT OF THE FFT ALGORITHM WITH RADIX-4

A radix-4 (base 4) algorithm can increase the execution speed of the FFT. FFT programs on higher radices and split radices have been developed. We use a decimation-in-frequency (DIF) decomposition process to introduce the development of the radix-4 FFT. The last or lowest decomposition of a radix-4 algorithm consists of four inputs and four outputs. The order or length of the FFT is 4^M, where M is the number of stages. For a 16-point FFT, there are only two stages or iterations, compared with four stages with the radix-2 algorithm. The DFT in (6.1) is decomposed into four summations, instead of two, as follows:

$$X(k) = \sum_{n=0}^{(N/4)-1} x(n)W^{nk} + \sum_{n=N/4}^{(N/2)-1} x(n)W^{nk} + \sum_{n=N/2}^{(3N/4)-1} x(n)W^{nk} + \sum_{n=3N/4}^{N-1} x(n)W^{nk} \quad (6.30)$$

Let $n = n + N/4, n = n + N/2, n = n + 3N/4$ in the second, third, and fourth summations, respectively. Then (6.30) can be written as

$$X(k) = \sum_{n=0}^{(N/4)-1} x(n)W^{nk} + W^{kN/4} \sum_{n=0}^{(N/4)-1} x(n+N/4)W^{nk}$$

$$+ W^{kN/2} \sum_{n=0}^{(N/4)-1} x(n+N/2)W^{nk} + W^{3kN/4} \sum_{n=0}^{(N/4)-1} x(n+3N/4)W^{nk} \quad (6.31)$$

which represents four $(N/4)$-point DFTs. Using

$$W^{kN/4} = \left(e^{-j2\pi/N}\right)^{kN/4} = e^{-jk\pi/2} = (-j)^k$$

$$W^{kN/2} = e^{-jk\pi} = (-1)^k$$

$$W^{3kN/4} = (j)^k$$

(6.31) becomes

$$X(k) = \sum_{n=0}^{(N/4)-1} \left[x(n) + (-j)^k x(n+N/4) + (-1)^k x(n+N/2) + (j)^k x(n+3N/4) \right] W^{nk} \quad (6.32)$$

Let $W_N^4 = W_{N/4}$. Equation (6.32) can be written as

$$X(4k) = \sum_{n=0}^{(N/4)-1} [x(n) + x(n+N/4) + x(n+N/2) + x(n+3N/4)] W_{N/4}^{nk} \quad (6.33)$$

$$X(4k+1) = \sum_{n=0}^{(N/4)-1} [x(n) - jx(n+N/4) - x(n+N/2) + jx(n+3N/4)] W_N^n W_{N/4}^{nk} \quad (6.34)$$

$$X(4k+2) = \sum_{n=0}^{(N/4)-1} [x(n) - x(n+N/4) + x(n+N/2) - x(n+3N/4)] W_N^{2n} W_{N/4}^{nk} \quad (6.35)$$

$$X(4k+3) = \sum_{n=0}^{(N/4)-1} [x(n) + jx(n+N/4) - x(n+N/2) - jx(n+3N/4)] W_N^{3n} W_{N/4}^{nk} \quad (6.36)$$

for $k = 0, 1, \ldots, (N/4) - 1$. Equations (6.33) through (6.36) represent a decomposition process yielding four four-point DFTs. The flow graph for a 16-point radix-4 decimation-in-frequency FFT is shown in Figure 6.11. Note the four-point butterfly in the flow graph. The $\pm j$ and -1 are not shown in Figure 6.11. The results shown in the flow graph are for the following exercise.

Exercise 6.4: 16-Point FFT with Radix-4

Given the input sequence $x(n)$ as in Exercise 6.2, representing a rectangular sequence $x(0) = x(1) = \ldots = x(7) = 1$, and $x(8) = x(9) = \ldots = x(15) = 0$. We will find the output sequence for a 16-point FFT with radix-4 using the flow graph in Figure 6.11. The twiddle constants are shown in Table 6.1.

TABLE 6.1   Twiddle Constants for 16-Point FFT with Radix-4

m	W_N^m	$W_{N/4}^m$
0	1	1
1	$0.9238 - j0.3826$	$-j$
2	$0.707 - j0.707$	-1
3	$0.3826 - j0.9238$	$+j$
4	$0 - j$	1
5	$-0.3826 - j0.9238$	$-j$
6	$-0.707 - j0.707$	-1
7	$-0.9238 - j0.3826$	$+j$

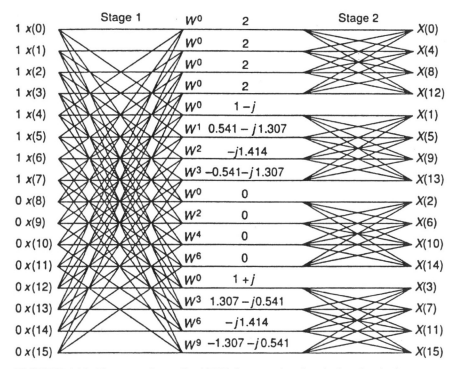

FIGURE 6.11. Sixteen-point radix-4 FFT flow graph using decimation-in-frequency.

The intermediate output sequence after stage 1 is shown in Figure 6.11. For example, after stage 1:

$$[x(0) + x(4) + x(8) + x(12)]W^0 = 1 + 1 + 0 + 0 = 2 \rightarrow x'(0)$$
$$[x(1) + x(5) + x(9) + x(13)]W^0 = 1 + 1 + 0 + 0 = 2 \rightarrow x'(1)$$
$$\vdots \qquad\qquad\qquad\qquad\qquad\qquad \vdots$$
$$[x(0) - jx(4) - x(8) + jx(12)]W^0 = 1 - j - 0 - 0 = 1 - j \rightarrow x'(4)$$
$$\vdots$$
$$[x(3) - x(7) + x(11) - x(15)]W^6 = 0 \rightarrow x'(11)$$
$$[x(0) + jx(4) - x(8) - jx(12)]W^0 = 1 + j - 0 - 0 = 1 + j \rightarrow x'(12)$$
$$\vdots \qquad\qquad\qquad\qquad\qquad\qquad \vdots$$
$$[x(3) + jx(7) - x(11) - jx(15)]W^9 = [1 + j - 0 - 0](-W^1)$$
$$= -1.307 - j0.541 \rightarrow x'(15)$$

For example, after stage 2:

$$X(3) = (1 + j) + (1.307 - j0.541) + (-j1.414) + (-1.307 - j0.541) = 1 - j1.496$$

and

$$X(15) = (1 + j)(1) + (1.307 - j0.541)(-j) + (-j1.414)(1)$$
$$+ (-1.307 - j0.541)(-j) = 1 + j5.028$$

The output sequence $X(0)$, $X(1)$, ..., $X(15)$ is identical to the output sequence obtained with the 16-point FFT with the radix-2 in Figure 6.6. These results also can be verified with MATLAB, described in Appendix D.

The output sequence is scrambled and needs to be resequenced or reordered. This can be done using a digit-reversal procedure, in a similar fashion as a bit reversal in a radix-2 algorithm. The radix-4 (base 4) uses the digits 0, 1, 2, 3. For example, the addresses of $X(8)$ and $X(2)$ need to be swapped because $(8)_{10}$ in base 10 or decimal is equal to $(20)_4$ in base 4. Digits 0 and 1 are reversed to yield $(02)_4$ in base 4, which is also $(02)_{10}$ in decimal.

Although mixed or higher radices can provide further reduction in computation, programming considerations become more complex. As a result, the radix-2 is still the most widely used, followed by the radix-4.

6.7 INVERSE FAST FOURIER TRANSFORM

The inverse discrete Fourier transform (IDFT) converts a frequency-domain sequence $X(k)$ into an equivalent sequence $x(n)$ in the time domain. It is defined as

$$x(n) = \frac{1}{N} \sum_{k=0}^{N-1} X(k) W^{-nk} \qquad n = 0, 1, \ldots, N - 1 \qquad (6.37)$$

Comparing (6.37) with the DFT equation definition in (6.1), we see that the FFT algorithm (forward) described previously can be used to find the IFFT (reverse), with the two following changes:

1. Adding a scaling factor of $1/N$
2. Replacing W^{nk} by its complex conjugate W^{-nk}

With the changes, the same FFT flow graphs can be used for the inverse fast Fourier transform (IFFT). We also develop programming examples to illustrate the inverse FFT.

A variant of the FFT, such as the fast Hartley transform (FHT), can be obtained readily from the FFT. Conversely, the FFT can be obtained from the FHT [10,11]. A development of the fast Hartley transform (FHT) with flow graphs and exercises for 8- and 16-point FHTs can be found in Ref. 12.

Exercise 6.5: Eight-Point IFFT

Let the output sequence $X(0) = 4$, $X(1) = 1 - j2.41$, ..., $X(7) = 1 + j2.41$ obtained in Exercise 6.1 become the input to an eight-point IFFT flow graph. Make the two changes (scaling and complex conjugate of W) to obtain an eight-point IFFT (reverse) flow graph from an eight-point FFT (forward) flow graph. The resulting flow graph becomes an IFFT flow graph similar to Figure 6.5. Verify that the resulting output sequence is $x(0) = 1, x(1) = 1, \ldots, x(7) = 0$, which represents the rectangular input sequence in Exercise 6.1.

6.8 PROGRAMMING EXAMPLES

Example 6.1: DFT of a Sequence of Real Numbers with Output from CCS Window (DFT)

This example illustrates the discrete Fourier transform (DFT) of an N-point sequence. Figure 6.12 shows a listing of the program *DFT.c*, which implements the DFT. The input sequence is x(n). The program calculates

$$X(k) = \text{DFT}\{x(n)\} = \sum_{n=0}^{N-1} x(n)W^{nk} \qquad k = 0, 1, \ldots, N-1$$

where $W = e^{-j2\pi/N}$ are the twiddle constants. This can be decomposed into a sum of real components and a sum of imaginary components, or

$$\text{Re}\{X(k)\} = \sum_{n=0}^{N-1} x(n)\cos(2\pi nk/N)$$

$$\text{IM}\{X(k)\} = \sum_{n=0}^{N-1} x(n)\sin(2\pi nk/N)$$

Using a sequence of real numbers with an integer number of cycles m, $X(k) = 0$ for all k, except at $k = m$ and at $k = N - m$.

Build this project as **DFT**. The input x(n) is a cosine with $N = 8$ data points. To test the results:

1. Select View → Watch Window and insert the two expressions *j* and *out* (right click on the Watch window). Click on +*out* to expand and view out[0] and out[1] that represent the real and imaginary components, respectively.

2. Place a breakpoint at the bracket "}" that follows the DFT function call.

3. Select Debug → Animate (Animation speed can be controlled through Options). Verify that the real component value out[0] is large (3996) at j = 1 and at j = 7, while small otherwise. Since x(n) is a one-cycle sequence, m = 1. Since the number of points is N = 8, a "spike" occurs at j = m = 1 and at j = N - m = 7. The following two MATLAB commands can be used to verify these results (see also Appendix D):

```
//DFT.c DFT of N-point from lookup table. Output from watch window

#include <stdio.h>
#include <math.h>
void dft(short *x, short k, int *out); //function prototype
#define N 8                            //number of data values
float pi = 3.1416;

short x[N] = {1000,707,0,-707,-1000,-707,0,707}; //1-cycle cosine

//short x[N]={0,602,974,974,602,0,-602,-974,-974,-602,
//            0,602,974,974,602,0,-602,-974,-974,-602}; //2-cycles sine

int out[2] = {0,0};                    //init Re and Im results

void dft(short *x, short k, int *out)  //DFT function
{
 int sumRe = 0;                        //init real component
 int sumIm = 0;                        //init imaginary component
 int i = 0;
 float cs = 0;                         //init cosine component
 float sn = 0;                         //init sine component

 for (i = 0; i < N; i++)               //for N-point DFT
    {
       cs = cos(2*pi*(k)*i/N);         //real component
       sn = sin(2*pi*(k)*i/N);         //imaginary component
       sumRe = sumRe + x[i]*cs;        //sum of real components
       sumIm = sumIm - x[i]*sn;        //sum of imaginary components
    }
 out[0] = sumRe;                       //sum of real components
 out[1] = sumIm;                       //sum of imaginary components
}
void main()
{
 int j;

 for (j = 0; j < N; j++)
    {
       dft(x,j,out);                   //call DFT function
    }
}
```

FIGURE 6.12. DFT implementation program with input from a lookup table (DFT.c).

x = [1000 707 0 -707 -1000 -707 0 707];

y = fft(x)

Note that the data values in the table are rounded (yielding a spike with a maximum value of 3996 in lieu of 4000). Since it is a cosine, the imaginary component out[1] is zero (small). In a real-time implementation, with $F_s = 8\,kHz$, the frequency generated would be at $f = F_s$ (number of cycles)/N $= 1\,kHz$.

4. Use a two-cycle sine data table with 20 points as input x(n). Within the program, change N to 20, comment the table that corresponds to the cosine (first input), and instead use the sine table values. Rebuild and Animate again. Verify a large negative value at j = 2 (-10,232) and a large positive value at j = N - m = 18 (10,232). For a real-time implementation, the magnitude of X(k), k = 0, 1, ... can be found. With $F_s = 8\,kHz$, the frequency generated would correspond to $f = 800\,Hz$.

Example 6.2: FFT of a Real-Time Input Signal Using an FFT Function in C (FFT256c)

Figure 6.13 shows a listing of the program *FFT256c.c*, which implements a 256-point FFT in real time, using an external input signal. It calls a generic FFT function in C, FFT.c (on the accompanying disk). This FFT function, used with the C31 DSK and the C30 EVM, is listed and described in Refs. 13 and 14.

The twiddle constants are generated within the program. The imaginary components of the input data are set to zero to illustrate this implementation. The magnitude of the resulting FFT (scaled) is taken for output to the codec. Three buffers are used:

1. *samples*: contains the data to be transformed
2. *iobuffer*: used to output a processed data as well as acquiring a new input sampled data
3. *x1*: contains the magnitude (scaled) of the tranformed (processed) data

On every sample period, an interrupt occurs. On each interrupt, an output value from a buffer (*iobuffer*) is sent to the codec's DAC and an input value is acquired and stored into the same buffer. An index (*buffercount*) to this buffer is used to set a flag when this buffer is full. When this buffer is full, it is copied to another buffer (*samples*), which will be used when calling the FFT function. The magnitude (scaled) of the processed FFT data, contained in a buffer x1, can now be copied to the I/O buffer, *iobuffer*, for output. In a filtering algorithm, processing can be done as each new sample is acquired. On the other hand, an FFT algorithm requires that an entire frame of data be available for processing.

```
//FFT256c.c FFT implementation calling C-coded FFT function

#include <math.h>
#define PTS 256                          //# of points for FFT
#define PI 3.14159265358979
typedef struct {float real,imag;}  COMPLEX;
void FFT(COMPLEX *Y, int n);        //FFT prototype
float iobuffer[PTS];                //as input and output buffer
float x1[PTS];                      //intermediate buffer
short i;                            //general purpose index variable
short buffercount = 0;              //number of new samples in iobuffer
short flag = 0;                     //set to 1 by ISR when iobuffer full
COMPLEX w[PTS];                     //twiddle constants stored in w
COMPLEX samples[PTS];               //primary working buffer

main()
{
 for (i = 0 ; i<PTS ; i++)          // set up twiddle constants in w
  {
   w[i].real = cos(2*PI*i/512.0);   //Re component of twiddle constants
   w[i].imag =-sin(2*PI*i/512.0);   //Im component of twiddle constants
  }
 comm_intr();                       //init DSK, codec, McBSP

 while(1)                           //infinite loop
  {
   while (flag == 0) ;              //wait until iobuffer is full
   flag = 0;                        //reset flag
   for (i = 0 ; i < PTS ; i++)      //swap buffers
    {
     samples[i].real=iobuffer[i];   //buffer with new data
     iobuffer[i] = x1[i];           //processed frame to iobuffer
    }
   for (i = 0 ; i < PTS ; i++)
     samples[i].imag = 0.0;         //imag components = 0

   FFT(samples,PTS);                //call function FFT.c

   for (i = 0 ; i < PTS ; i++)      //compute magnitude
    {
     x1[i] = sqrt(samples[i].real*samples[i].real
          + samples[i].imag*samples[i].imag)/32;
    }
   x1[0] = 32000.0;                 //negative spike(with AD535)for ref
  }                                 //end of infinite loop
}                                   //end of main

interrupt void c_int11()           //ISR
{
  output_sample((int)(iobuffer[buffercount]));       //out from iobuffer
  iobuffer[buffercount++]=(float)(input_sample());   //input to iobuffer
  if (buffercount >= PTS)                            //if iobuffer full
  {
      buffercount = 0;                                //reinit buffercount
      flag = 1;                                       //set flag
  }
}
```

FIGURE 6.13. FFT program of real-time input calling a C-coded FFT function (FFT256c.c).

202

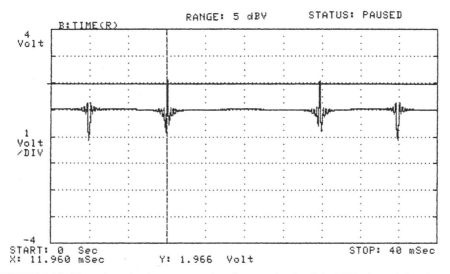

FIGURE 6.14. Time-domain plot representing the magnitude of the FFT of a real-time input.

Build and run this project as **FFT256c**. Input a 2-kHz sine wave with an amplitude of approximately 0.5 to 1 V p-p. Figure 6.14 shows a time-domain representation of the magnitude of the transformed data, obtained with an HP dynamic signal analyzer (you can use an oscilloscope). The two negative spikes are $256(T_s) = 32\,\text{ms}$ apart, as shown in Figure 6.14. This interval also represents the sampling frequency F_s. The location of the first positive spike then corresponds to a frequency of 2 kHz (the mid-distance between the two spikes corresponds to 4 kHz). The location of the second positive spike corresponds to the folding frequency of $F_s - f = 6\,\text{kHz}$. Increase the frequency of the input signal and observe the convergence of the two spikes toward the 4-kHz Nyquist frequency.

Example 6.3: FFT of a Sinusoidal Signal from a Table Using TI's C Callable FFT Function (FFTsinetable)

Figure 6.15 shows a listing of the program *FFTsinetable.c*, which illustrates a C program calling TI's floating-point FFT function *cfftr2_dit.sa*, available at TI's Web site (also on disk). The twiddle constants are calculated within the program. The imaginary components of the twiddle constants are negated, as required (assumed) by the FFT function. The FFT function also assumes $N/2$ complex twiddle constants. It is important to align the data in memory (on an 8-byte boundary). Both the input data and the twiddle constants are structured as "complex."

The input signal consists of sine data values set in a table as real input data. The imaginary components of the input sine data are set to zero. The input data are arranged in memory as successive real and imaginary number pairs, as required (assumed) by the FFT function. The resulting ouput is still complex.

```
//FFTsinetable.c FFT{sine}from table. Calls TI float-point FFT function

#include <math.h>
#define N 32                           //number of FFT points
#define SQRT_N 32                      //SQRT_N >= SQRT(N)
#define FREQ 8                         //# of points/cycle
#define RADIX 2                        //radix or base
#define DELTA (2*PI)/N                 //argument for sine/cosine
#define TAB_PTS 32                     //# of points in sine_table
#define PI 3.14159265358979
short i = 0;
short iTwid[SQRT_N];                   //N/2 + 1 > sqrt(N)
short iData[N];                        //index for bitrev X
float Xmag[N];                         //magnitude spectrum of x
typedef struct Complex_tag {float re,im;}Complex;
Complex W[N/RADIX];                    //array for twiddle constants
Complex x[N];                          //N complex data values
#pragma DATA_ALIGN(W,sizeof(Complex))  //align boundary size complex

short sine_table[TAB_PTS] = {0,195,383,556,707,831,924,981,1000,
981,924,831,707,556,383,195,-0,-195,-383,-556,-707,-831,-924,-981,
-1000,-981,-924,-831,-707,-556,-383,-195};

void main()
{
 for( i = 0 ; i < N/RADIX ; i++ )
  {
   W[i].re = cos(DELTA*i);             //real component of W
   W[i].im = sin(DELTA*i);             //neg imag component
  }                                    //see cfftr2_dit
 for( i = 0 ; i < N ; i++ )
  {
   x[i].re=3*sine_table[FREQ*i % TAB_PTS]; //wrap when i=TAB_PTS
   x[i].im = 0 ;                       //zero imaginary part
  }
 digitrev_index(iTwid, N/RADIX, RADIX); //produces index for bitrev() W
 bitrev(W, iTwid, N/RADIX);            //bit reverse W

 cfftr2_dit(x, W, N ) ;               //TI floating-pt complex FFT

 digitrev_index(iData, N, RADIX);     //produces index for bitrev() X
 bitrev(x, iData, N);                 //freq scrambled->bit-reverse X
 for(i = 0 ; i < N ; i++ )
    Xmag[i] = sqrt(x[i].re*x[i].re+x[i].im*x[i].im ); //magnitude of X

 comm_poll( ) ;                       //init DSK,codec,McBSP
 while (1)                            //infinite loop
  {
   output_sample(32000) ;             //negative spike as reference
   for (i = 1; i < N; i++)
     output_sample((int)Xmag[i]);     //output magnitude samples
  }
}
```

FIGURE 6.15. FFT program of input data from a table using TI's optimized floating-point complex FFT function (FFTsinetable.c).

The FFT function *cfftr2_dit.sa* uses a decimation-in-time, radix 2, and takes the FFT of a "complex" input signal. Two support functions, *digitrev_index.c* and *bitrev.sa*, are used in conjunction with the complex FFT function for bit reversal. These two support files are also available through TI's Web site (also on disk). The FFT function *cfftr2_dit.sa* assumes that the input data item x is in normal order while the FFT coefficients or twiddle constants are in reverse order. As a result, the support function *digitrev_index.c*, to produce the index for bit reversal, and *bitrev.sa*, to perform the bit reversal on the twiddle constants, are called before the FFT function is invoked. These two support files for bit reversal are again called to bit-reverse the resulting scrambled output.

N is the number of complex input (note that the input data consist of $2N$ elements) or output data, so that an N-point FFT is performed. SQRT_N is used by the bit-reversal support functions. FREQ determines the frequency of the input sine data by selecting the number of points per cycle within the data table. With FREQ set at 8, every eighth point from the table is selected, starting with the first data point. The modulo operator is used as a flag to reinitialize the index. The following four points (scaled) within one period are selected: 0, 1000, 0, and −1000. Example 2.4 (sine2sliders) illustrates this indexing scheme to select different number of data points within a table.

The magnitude of the resulting FFT is taken. The line of code

```
output_sample(32000);
```

outputs a negative spike of approximately −1.1 V (not positive, due to the 2's-complement format of the AD535 codec and a dc offset of approximately 1.1 V). It is used as a reference scheme. The input data are scaled so that the output magnitude is positive (again due to the codec data format). The sampling rate is achieved through polling.

Build and run this project as **FFTsinetable**. The two support files for bit reversal and the complex FFT function also are included in the Source project. Figure 6.16 shows a time-domain plot of the resulting output (obtained with an HP dynamic signal analyzer). Since an output occurs every T_s, the time interval for 32 points corresponds to $32T_s$, or 32(0.125 ms) = 4 ms. A negative spike is then repeated every 4 ms. This provides a reference, since the time interval between the two negative spikes corresponds to the sampling frequency of 8 kHz. The center of this time interval then corresponds to the Nyquist frequency of 4 kHz (2 ms from the negative spike). The first positive spike occurs at 1 ms from the first negative spike. This corresponds to a frequency of $f = F_s/4 = 2$ kHz. The second positive spike occurs at 3 ms and corresponds to the folding frequency of $F_s - f = 6$ kHz.

Change FREQ to 4 in order to select eight sine data values within the table. Verify that the output is a 1-kHz signal (obtain a plot similar to that in Figure 6.14 from an oscilloscope). A FREQ value of 12 produces an output of 3 kHz. A FREQ value of 15 shows the two positive spikes at the center (between the two negative spikes). Note that aliasing occurs for frequencies larger than 4 kHz. To illustrate that,

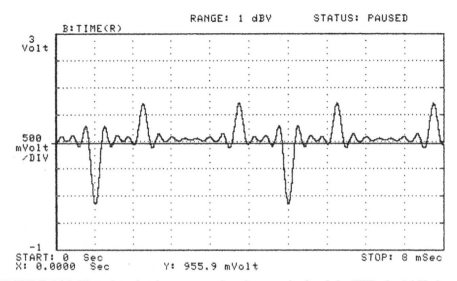

FIGURE 6.16. Time-domain plot representing the magnitude of the FFT of a 2-kHz input data from a table obtained using TI's FFT function.

change `FREQ` to a value of 20. Verify that the output is an aliased signal at 3 kHz, in lieu of 5 kHz. A `FREQ` value of 24 would show an aliased signal of 2 kHz in lieu of 6 kHz.

The number of cycles is documented within the function $cfftr2\_dit.sa$ (by TI) as

```
Cycles = ((2N) + 23)log₂(N) + 6
```

For a 1024-point FFT, the number of cycles would be $(2071)(10) + 6 = 20,716$. This corresponds to a time of $t = 20,716$ cycles/(150 MHz) = 138 µs.

6.8.1 Fast Convolution

The following examples show how the FFT enables signals to be processed in the frequency domain. Fast convolution [19,20] takes less computational effort and is potentially more accurate than time-domain implementation of FIR filters having very large numbers of coefficients.

Example 6.4: Fast Convolution with Overlap-Add for FIR Implementation Using TI's Floating-Point FFT Functions (`fastconvo`)

Figure 6.17 shows a listing of the program `fastconvo.c` to implement an FIR filter and illustrate the fast convolution's overlap-add scheme [19,20]. TI's floating-point

FFT support functions, bitrev.sa, digitrev_index.c, and cfftr2_dit.sa, were introduced in Example 6.3. In addition, TI's inverse complex FFT function icfftr2_dif.sa (radix-2, DIF) is used here. This function expects its input to be scrambled or to be in bit-reversed order. As a result, the bit-reversed output of the complex FFT function cfftr2_dit.sa need not be reordered, and the support files for bit reversal, digitrev_index.c and bitrev.sa, are not needed after the FFT section of the program. Both data (samples) and filter coefficients (h) are in bit-reversed order and may be multiplied together in that order.

Build this project as **Fastconvo** (use compiler optimization level -o1). The time-domain filter coefficients are read from the file coeffs.h. Verify that the output yields a 2-kHz bandpass filter. The filter coefficients are the same as BP55.cof, with a center frequency at $F_s/4$, introduced in Example 4.4.

The coefficient file coeffs.h also contains a set of coefficients identical to LP55.cof, which represents a lowpass FIR filter with a cutoff frequency at $F_s/8$, also introduced in Example 4.4. Edit the file coeffs.h to implement/verify this lowpass filter.

Several buffers are used, and iobuffer is the primary input/output buffer. At each sampling interval, the ISR is executed. The next output value is read from iobuffer, output to the codec, and then replaced by a new input sample. After PTS/2 sampling instants, iobuffer contains a new frame of PTS/2 input samples. This situation is signaled by setting flag to 1.

The main program waits for this flag signal using

```
while (flag == 0);
```

and subsequently carries out the following operations:

1. Resets flag to zero
2. Copies the contents of the buffer iobuffer (frame of new input samples) to the first PTS/2 locations of the buffer samples
3. Copies the contents of the buffer overlap (previously computed frame of output samples) to the buffer iobuffer
4. Processes the new frame of input samples to compute the next frame of output samples

The frame processing operation (within an infinite loop) has PTS/2 sampling periods in which to execute and comprises the following steps:

1. The contents of the last PTS/2 locations of the samples buffer (real parts) are copied to the overlap buffer. These time-domain data may be thought of as the overlapping latter-half (PTS/2 samples) of the *previous* frame processing operation.

```
//FastConvo.c FIR filter implemented using overlap-add fast convolution

#include <math.h>
#include "coeffs.h"                  //time domain FIR coefficients
#define PI 3.14159265358979
#define PTS 256                      //number of points for FFT
#define SQRT_PTS 16                  //used in twiddle factor calc.
#define RADIX 2                      //passed to TI FFT routines
#define DELTA (2*PI)/PTS
typedef struct Complex_tag {float real, imag;} COMPLEX ;
#pragma DATA_ALIGN(W, sizeof(COMPLEX))
#pragma DATA_ALIGN(samples, sizeof(COMPLEX))
#pragma DATA_ALIGN(h, sizeof(COMPLEX))
COMPLEX W[PTS/RADIX] ;               //twiddle factor array
COMPLEX samples[PTS];                //processing buffer
COMPLEX h[PTS];                      //FIR filter coefficients
short buffercount = 0;               //buffer count for iobuffer samples
float iobuffer[PTS/2];               //primary input/output buffer
float overlap[PTS/2];                //intermediate result buffer
short i;                             //index variable
short flag = 0;                      //set to indicate iobuffer full
float a, b;                          //variables used in complex multiply
short NUMCOEFFS = sizeof(coeffs)/sizeof(float);
short iTwid[SQRT_PTS] ;              //PTS/2 + 1 > sqrt(PTS)

interrupt void c_int11(void)         //ISR
{
 output_sample((int)(iobuffer[buffercount]));
 iobuffer[buffercount++] = (float)(input_sample());
 if (buffercount >= PTS/2)           //for overlap-add method iobuffer
  {                                  //is half size of FFT used
   buffercount = 0;
   flag = 1;
  }
}


main()
{                                    //set up array of twiddle factors
 digitrev_index(iTwid, PTS/RADIX, RADIX);
 for(i = 0 ; i < PTS/RADIX ; i++)
 {
  W[i].real = cos(DELTA*i);
  W[i].imag = sin(DELTA*i);
 }
```

FIGURE 6.17. Fast convolution program using overlap-add with TI's floating-point FFT functions (fastconvo.c).

```
bitrev(W, iTwid, PTS/RADIX);        //bit reverse order W
for (i = 0 ; i<PTS ; i++)           //initialise PTS element
 {                                   //of COMPLEX to hold real-valued
  h[i].real = 0.0;                   //time domain FIR filter coefficients
  h[i].imag = 0.0;
 }
for (i = 0 ; i < NUMCOEFFS ; i++)
 {                                   //read FIR filter coeffs
  h[i].real = coeffs[i];             //NUMCOEFFS should be less than PTS/2
 }
cfftr2_dit(h,W,PTS);                 //transform filter coeffs
comm_intr();                         //initialise DSK, codec, McBSP
while(1)                             //frame processing infinite loop
 {
   while (flag == 0);                //wait for iobuffer full
    flag = 0;
    for (i = 0 ; i<PTS/2 ; i++) //iobuffer into first half of
    {                                //samples buffer
     samples[i].real = iobuffer[i];
     iobuffer[i] = overlap[i];       //previously processed output
    }                                //to iobuffer
   for (i = 0 ; i<PTS/2 ; i++)
   {                                 //second half of samples to overlap
    overlap[i] = samples[i+PTS/2].real;
    samples[i+PTS/2].real = 0.0; //zero-pad input from iobuffer
   }
   for (i=0; i<PTS ; i++)
    samples[i].imag = 0.0;           //init imag parts in samples buffer
   cfftr2_dit(samples,W,PTS);        //complex FFT function from TI

   for (i=0 ; i<PTS ; i++)           //frequency-domain representation
   {                                 //complex multiply samples by h
    a = samples[i].real;
    b = samples[i].imag;
    samples[i].real = h[i].real*a - h[i].imag*b;
    samples[i].imag = h[i].real*b + h[i].imag*a;
   }

   icfftr2_dif(samples,W,PTS);       //inverse FFT function from TI

   for (i=0 ; i<PTS ; i++)
    samples[i].real /= PTS;
   for (i=0 ; i<PTS/2 ; i++)         //add first half of samples
    overlap[i] += samples[i].real;    //to overlap
 }                                   //end of while(1)
}                                    //end of main()
```

FIGURE 6.17. (*Continued*)

2. The last PTS/2 locations of the buffer samples are zero-padded. The buffer samples now contains PTS/2 new samples followed by PTS/2 zeros.

3. The buffer `samples` is transformed in-place into the frequency domain using a PTS-point FFT.

4. The complex frequency-domain sample values are multiplied by the complex frequency-domain filter coefficients stored in `h`.

5. The results are transformed back into the time domain by applying a PTS-point IFFT to the contents of the samples buffer. The resulting PTS time-domain samples will be real-valued.

6. The contents of the first PTS/2 locations of the buffer `samples` (i.e., the former-half of the *current* frame processing result) are added to the contents of the overlap buffer.

Since the input and output signals are real-valued, so are the buffers `iobuffer` and `overlap`. However, since the frequency-domain representation of these signals is complex, the buffer `samples` and the array of filter coefficients `h` are complex, requiring two floating-point values (real and imaginary parts) per sample.

A faster and more efficient implementation of buffering is possible using pointers rather than copying data from one buffer to another, but the latter approach is adopted for purposes of clarity.

Alternative Version for Simulation

The program ***fastconvosim.c*** (on the accompanying disk) is a non-real-time version of the program ***fastconvo.c***, which processes a prestored sequence of input samples. Using breakpoints (locations specified within the program), the user can step through the various stages in the overlap-add process, viewing the contents of each of the buffers at each step. Figure 6.18 shows a typical view of the contents of the buffers (obtained with CCS): ***iobuffer***, ***h***, ***samples***, and ***overlap***, at an intermediate stage in the process.

Example 6.5: Graphic Equalizer (`graphicEQ`)

Figure 6.19 shows a listing of the program `graphicEQ.c`, which implements a three-band graphic equalizer. TI's floating-point complex FFT and inverse FFT support functions are used again in this project (see also Examples 6.3 and 6.4).

`graphicEQcoeff.h` contains three sets of coefficients; lowpass at 1.3 kHz, bandpass between 1.3 and 2.6 kHz, and highpass at 2.6 kHz, designed with MATLAB's function fir1. Both the input samples and the three sets of coefficients are transformed into the frequency domain. The filtering is performed in the frequency domain based on the overlap-add scheme used in Example 6.4 [19,20]. Note

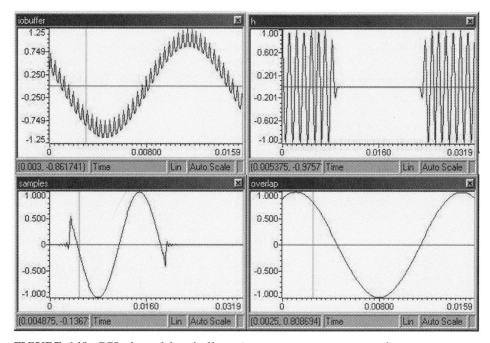

FIGURE 6.18. CCS plots of four buffers—iobuffer, h, samples, and overlap—at an intermediate processing stage using the simulation version program fastconvosim.c (on the disk).

that the complex multiplication (H)(X), where H represents the transfer function and X the input sample, yields

$$(H_R + jH_I)(X_R + jX_I) = (H_R X_R - H_I X_I) + j\,(H_R X_I + H_I X_R)$$

as used in the program, where $j = \sqrt{-1}$

ISR continuously (every sample period T_s) outputs a value from the buffer iobuffer, then inputs a new value until iobuffer is full. At such time a new frame of input data is available. The iobuffer index is initialized and the flag is set. The main program waits for this flag to be set, then resets it.

Build this project as **graphicEQ** (use the optimization level −o1). Test this project using an input voice file such as TheForce.wav (see Example 4.9) or noise. Verify that the low- and high-frequency components are accentuated while the midrange frequency components are attenuated. This is because the filter coefficients are scaled in the program by bass_gain and treble_gain, initially set to 1, and by mid_gain, initially set to zero. The slider file graphicEQ.gel (on the disk) allows you to control the three frequency bands independently. Figure 6.20 shows the output spectrum obtained with a signal analyzer using noise as input and two different gain settings.

```
//GraphicEQ.c Graphic Equalizer using TI floating-point FFT functions

#include <math.h>
#include "GraphicEQcoeff.h"          //time-domain FIR coefficients
#define PI 3.14159265358979
#define PTS 256                       //number of points for FFT
#define SQRT_PTS 16
#define RADIX 2
#define DELTA (2*PI)/PTS
typedef struct Complex_tag {float real,imag;} COMPLEX;
#pragma DATA_ALIGN(W,sizeof(COMPLEX))
#pragma DATA_ALIGN(samples,sizeof(COMPLEX))
#pragma DATA_ALIGN(h,sizeof(COMPLEX))
COMPLEX W[PTS/RADIX] ;                //twiddle array
COMPLEX samples[PTS];
COMPLEX h[PTS];
COMPLEX bass[PTS], mid[PTS], treble[PTS];
short buffercount = 0;                //buffer count for iobuffer samples
float iobuffer[PTS/2];                //primary input/output buffer
float overlap[PTS/2];                 //intermediate result buffer
short i;                              //index variable
short flag = 0;                       //set to indicate iobuffer full
float a, b;                           //variables for complex multiply
short NUMCOEFFS = sizeof(lpcoeff)/sizeof(float);
short iTwid[SQRT_PTS] ;               //PTS/2+1 > sqrt(PTS)
float bass_gain = 1.0;                //initial gain values
float mid_gain = 0.0;                 //change with GraphicEQ.gel
float treble_gain = 1.0;

interrupt void c_int11(void)          //ISR
{
 output_sample((int)(iobuffer[buffercount]));
 iobuffer[buffercount++] = (float)(input_sample());
 if (buffercount >= PTS/2)            //for overlap-add method iobuffer
  {                                   //is half size of FFT used
   buffercount = 0;
   flag = 1;
  }
}

main()
{
 digitrev_index(iTwid, PTS/RADIX, RADIX);
 for( i = 0; i < PTS/RADIX; i++ )
  {
   W[i].real = cos(DELTA*i);
   W[i].imag = sin(DELTA*i);
  }
 bitrev(W, iTwid, PTS/RADIX);         //bit reverse W

 for (i=0 ; i<PTS ; i++)
  {
   bass[i].real = 0.0;
   bass[i].imag = 0.0;
   mid[i].real = 0.0;
   mid[i].imag = 0.0;
   treble[i].real = 0.0;
   treble[i].imag = 0.0;
  }
```

FIGURE 6.19. Equalizer program using TI's floating-point FFT functions (graphicEQ.c).

```
for (i=0; i<NUMCOEFFS; i++)      //same # of coeff for each filter
  {
   bass[i].real = lpcoeff[i];    //lowpass coeff
   mid[i].real = bpcoeff[i];     //bandpass coeff
   treble[i].real = hpcoeff[i];  //highpass coef
  }

cfftr2_dit(bass,W,PTS);          //transform each band into frequency
cfftr2_dit(mid,W,PTS);
cfftr2_dit(treble,W,PTS);


comm_intr();                     //initialise DSK, codec, McBSP
while(1)                         //frame processing infinite loop
  {
   while (flag == 0);            //wait for iobuffer full
         flag = 0;
   for (i=0 ; i<PTS/2 ; i++)     //iobuffer into samples buffer
     {
      samples[i].real = iobuffer[i];
      iobuffer[i] = overlap[i];  //previously processed output
     }                           //to iobuffer
   for (i=0 ; i<PTS/2 ; i++)
     {                           //upper-half samples to overlap
      overlap[i] = samples[i+PTS/2].real;
      samples[i+PTS/2].real = 0.0; //zero-pad input from iobuffer
     }
   for (i=0 ; i<PTS ; i++)
     samples[i].imag = 0.0;      //init samples buffer

   cfftr2_dit(samples,W,PTS);

   for (i=0 ; i<PTS ; i++)       //construct freq domain filter
     {                           //sum of bass,mid,treble coeffs
      h[i].real = bass[i].real*bass_gain + mid[i].real*mid_gain
            + treble[i].real*treble_gain;
      h[i].imag = bass[i].imag*bass_gain + mid[i].imag*mid_gain
            + treble[i].imag*treble_gain;
     }
   for (i=0; i<PTS; i++)         //frequency-domain representation
     {                           //complex multiply samples by h
      a = samples[i].real;
      b = samples[i].imag;
      samples[i].real = h[i].real*a - h[i].imag*b;
      samples[i].imag = h[i].real*b + h[i].imag*a;
     }

   icfftr2_dif(samples,W,PTS);

   for (i=0 ; i<PTS ; i++)
     samples[i].real /= PTS;
   for (i=0 ; i<PTS/2 ; i++)     //add 1st half to overlap
     overlap[i] += samples[i].real;
  }                              //end of infinite loop
}                                //end of main()
```

FIGURE 6.19. (*Continued*)

213

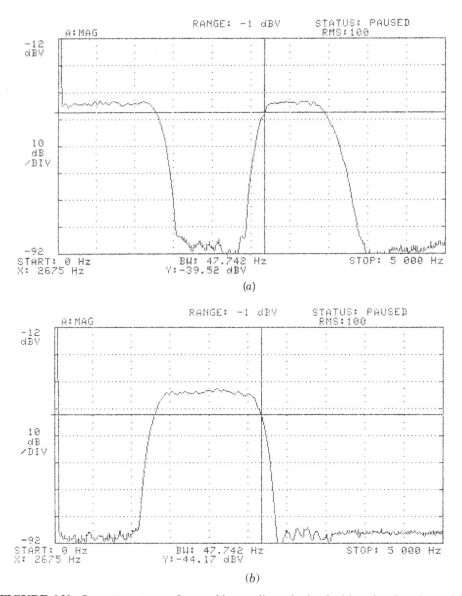

FIGURE 6.20. Output spectrum of a graphic equalizer obtained with a signal analyzer: (*a*) bass_gain = treble_gain = 1, mid_gain = 0; (*b*) bass_gain = treble_gain = 0, mid_gain = 1.

REFERENCES

1. J. W. Cooley and J. W. Tukey, An algorithm for the machine calculation of complex Fourier series, *Mathematics of Computation*, Vol. 19, 1965, pp. 297–301.

2. J. W. Cooley, How the FFT gained acceptance, *IEEE Signal Processing*, Jan. 1992, pp. 10–13.

3. J. W. Cooley, The structure of FFT and convolution algorithms, from a tutorial, *IEEE 1990 International Conference on Acoustics, Speech, and Signal Processing*, Apr. 1990.

4. C. S. Burrus and T. W. Parks, *DFT/FFT and Convolution Algorithms: Theory and Implementation*, Wiley, New York, 1988.

5. G. D. Bergland, A guided tour of the fast Fourier transform, *IEEE Spectrum*, Vol. 6, 1969, pp. 41–51.

6. E. O. Brigham, *The Fast Fourier Transform*, Prentice Hall, Upper Saddle River, NJ, 1974.

7. S. Winograd, On computing the discrete Fourier transform, *Mathematics of Computation*, Vol. 32, 1978, pp. 175–199.

8. H. F. Silverman, An introduction to programming the Winograd Fourier transform algorithm (WFTA), *IEEE Transactions on Acoustics, Speech, and Signal Processing*, Vol. ASSP-25, Apr. 1977, pp. 152–165.

9. P. E. Papamichalis, ed., *Digital Signal Processing Applications with the TMS320 Family: Theory, Algorithms, and Implementations*, Vol. 3, Texas Instruments, Dallas, TX, 1990.

10. R. N. Bracewell, Assessing the Hartley transform, *IEEE Transactions on Acoustics, Speech, and Signal Processing*, Vol. ASSP-38, 1990, pp. 2174–2176.

11. R. N. Bracewell, *The Hartley Transform*, Oxford University Press, New York, 1986.

12. H. V. Sorensen, D. L. Jones, M. T. Heidman, and C. S. Burrus, Real-valued fast Fourier transform algorithms, *IEEE Transactions on Acoustics, Speech, and Signal Processing*, Vol. ASSP-35, 1987, pp. 849–863.

13. R. Chassaing, *Digital Signal Processing Laboratory Experiments Using C and the TMS320C31 DSK*, Wiley, New York, 1999.

14. R. Chassaing, *Digital Signal Processing with C and the TMS320C30*, Wiley, New York, 1992.

15. P. M. Embree and B. Kimble, *C Language Algorithms for Digital Signal Processing*, Prentice Hall, Upper Saddle River, NJ, 1990.

16. S. Kay and R. Sudhaker, A zero crossing spectrum analyzer, *IEEE Transactions on Acoustics, Speech, and Signal Processing*, Vol. ASSP-34, Feb. 1986, pp. 96–104.

17. P. Kraniauskas, A plain man's guide to the FFT, *IEEE Signal Processing*, Apr. 1994.

18. J. R. Deller, Jr., Tom, Dick, and Mary discover the DFT, *IEEE Signal Processing*, Apr. 1994.

19. A. V. Oppenheim and R. Schafer, *Discrete-Time Signal Processing*, Prentice Hall, Upper Saddle River, NJ, 1989.

20. J. G. Proakis and D. G. Manolakis, *Digital Signal Processing*, Upper Saddle River, NJ, 1996.

7

Adaptive Filters

- Adaptive structures
- The least mean squares (LMS) algorithm
- Programming examples for noise cancellation and system identification using C code

Adaptive filters are best used in cases where signal conditions or system parameters are slowly changing and the filter is to be adjusted to compensate for this change. The least mean squares (LMS) criterion is a search algorithm that can be used to provide the strategy for adjusting the filter coefficients. Programming examples are included to give a basic intuitive understanding of adaptive filters.

7.1 INTRODUCTION

In conventional FIR and IIR digital filters, it is assumed that the process parameters to determine the filter characteristics are known. They may vary with time, but the nature of the variation is assumed to be known. In many practical problems, there may be a large uncertainty in some parameters because of inadequate prior test data about the process. Some parameters might be expected to change with time, but the exact nature of the change is not predictable. In such cases it is highly desirable to design the filter to be self-learning, so that it can adapt itself to the situation at hand.

The coefficients of an adaptive filter are adjusted to compensate for changes in input signal, output signal, or system parameters. Instead of being rigid, an adaptive system can learn the signal characteristics and track slow changes. An adaptive filter can be very useful when there is uncertainty about the characteristics of a signal or when these characteristics change.

216

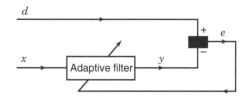

FIGURE 7.1. Basic adaptive filter structure.

Figure 7.1 shows a basic adaptive filter structure in which the adaptive filter's output y is compared with a desired signal d to yield an error signal e, which is fed back to the adaptive filter. The coefficients of the adaptive filter are adjusted, or optimized, using a least mean squares (LMS) algorithm based on the error signal.

We discuss here only the LMS searching algorithm with a linear combiner (FIR filter), although there are several strategies for performing adaptive filtering. The output of the adaptive filter in Figure 7.1 is

$$y(n) = \sum_{k=0}^{N-1} w_k(n)x(n-k) \tag{7.1}$$

where $w_k(n)$ represent N weights or coefficients for a specific time n. The convolution equation (7.1) was implemented in Chapter 4 in conjunction with FIR filtering. It is common practice to use the terminology of weights w for the coefficients associated with topics in adaptive filtering and neural networks.

A performance measure is needed to determine how good the filter is. This measure is based on the error signal,

$$e(n) = d(n) - y(n) \tag{7.2}$$

which is the difference between the desired signal $d(n)$ and the adaptive filter's output $y(n)$. The weights or coefficients $w_k(n)$ are adjusted such that a mean squared error function is minimized. This mean squared error function is $E[e^2(n)]$, where E represents the expected value. Since there are k weights or coefficients, a gradient of the mean squared error function is required. An estimate can be found instead using the gradient of $e^2(n)$, yielding

$$w_k(n+1) = w_k(n) + 2\beta e(n)x(n-k) \qquad k = 0, 1, \ldots, N-1 \tag{7.3}$$

which represents the LMS algorithm [1–3]. Equation (7.3) provides a simple but powerful and efficient means of updating the weights, or coefficients, without the need for averaging or differentiating, and will be used for implementing adaptive filters. The input to the adaptive filter is $x(n)$, and the rate of convergence and accuracy of the adaptation process (adaptive step size) is β.

For each specific time n, each coefficient, or weight, $w_k(n)$ is updated or replaced by a new coefficient, based on (7.3), unless the error signal $e(n)$ is zero. After the filter's output $y(n)$, the error signal $e(n)$ and each of the coefficients $w_k(n)$ are updated for a specific time n, a new sample is acquired (from an ADC) and the adaptation process is repeated for a different time. Note that from (7.3), the weights are not updated when $e(n)$ becomes zero.

The linear adaptive combiner is one of the most useful adaptive filter structures and is an adjustable FIR filter. Whereas the coefficients of the frequency-selective FIR filter discussed in Chapter 4 are fixed, the coefficients, or weights, of the adaptive FIR filter can be adjusted based on a changing environment such as an input signal. Adaptive IIR filters (not discussed here) can also be used. A major problem with an adaptive IIR filter is that its poles may be updated during the adaptation process to values outside the unit circle, making the filter unstable.

The programming examples developed later will make use of equations (7.1)–(7.3). In (7.3) we simply use the variable β in lieu of 2β.

7.2 ADAPTIVE STRUCTURES

A number of adaptive structures have been used for different applications in adaptive filtering.

1. *For noise cancellation.* Figure 7.2 shows the adaptive structure in Figure 7.1 modified for a noise cancellation application. The desired signal d is corrupted by uncorrelated additive noise n. The input to the adaptive filter is a noise n' that is correlated with the noise n. The noise n' could come from the same source as n but modified by the environment. The adaptive filter's output y is adapted to the noise n. When this happens, the error signal approaches the desired signal d. The overall output is this error signal and not the adaptive filter's output y. This structure will be further illustrated with programming examples using C code.

2. *For system identification.* Figure 7.3 shows an adaptive filter structure that can be used for system identification or modeling. The same input is to an unknown system in parallel with an adaptive filter. The error signal e is the difference between the response of the unknown system d and the response of the adaptive filter y. This error signal is fed back to the adaptive filter and

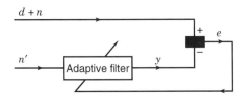

FIGURE 7.2. Adaptive filter structure for noise cancellation.

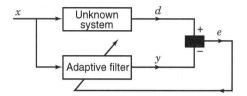

FIGURE 7.3. Adaptive filter structure for system identification.

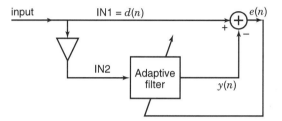

FIGURE 7.4. Adaptive predictor structure.

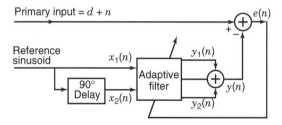

FIGURE 7.5. Adaptive notch structure with two weights.

is used to update the adaptive filter's coefficients until the overall output $y = d$. When this happens, the adaptation process is finished, and e approaches zero. In this scheme, the adaptive filter models the unknown system. This structure is illustrated later with three programming examples.

3. *Adaptive predictor.* Figure 7.4 shows an adaptive predictor structure which can provide an estimate of an input. This structure is illustrated later with a programming example.

4. Additional structures have been implemented, such as:

   (a) *Notch with two weights*, which can be used to notch or cancel/reduce a sinusoidal noise signal. This structure has only two weights or coefficients. This structure is shown in Figure 7.5 and is illustrated in Refs. 1, 3, and 4 using the C31 processor.

   (b) Adaptive channel equalization, used in a modem to reduce channel distortion resulting from the high speed of data transmission over telephone channels.

The LMS is well suited for a number of applications, including adaptive echo and noise cancellation, equalization, and prediction.

Other variants of the LMS algorithm have been employed, such as the sign-error LMS, the sign-data LMS, and the sign-sign LMS.

1. For the sign-error LMS algorithm, (7.3) becomes

$$w_k(n+1) = w_k(n) + \beta \, \mathrm{sgn}[e(n)]x(n-k) \qquad (7.4)$$

where sgn is the signum function,

$$\mathrm{sgn}(u) = \begin{cases} 1 & \text{if } u \geqslant 0 \\ -1 & \text{if } u < 0 \end{cases} \qquad (7.5)$$

2. For the sign-data LMS algorithm, (7.3) becomes

$$w_k(n+1) = w_k(n) + \beta e(n) \, \mathrm{sgn}[x(n-k)] \qquad (7.6)$$

3. For the sign-sign LMS algorithm, (7.3) becomes

$$w_k(n+1) = w_k(n) + \beta \, \mathrm{sgn}[e(n)] \, \mathrm{sgn}[x(n-k)] \qquad (7.7)$$

which reduces to

$$w_k(n+1) = \begin{cases} w_k(n) + \beta & \text{if } \mathrm{sgn}[e(n)] = \mathrm{sgn}[x(n-k)] \\ w_k(n) - \beta & \text{otherwise} \end{cases} \qquad (7.8)$$

which is more concise from a mathematical viewpoint because no multiplication operation is required for this algorithm.

The implementation of these variants does not exploit the pipeline features of the TMS320C6x processor. The execution speed on the TMS320C6x for these variants can be slower than for the basic LMS algorithm, due to additional decision-type instructions required for testing conditions involving the sign of the error signal or the data sample.

The LMS algorithm has been quite useful in adaptive equalizers, telephone cancelers, and so forth. Other methods, such as the recursive least squares (RLS) algorithm [4], can offer faster convergence than the basic LMS but at the expense of more computations. The RLS is based on starting with the optimal solution and then using each input sample to update the impulse response in order to maintain that optimality. The right step size and direction are defined over each time sample.

Adaptive algorithms for restoring signal properties can also be found in Ref. 4. Such algorithms become useful when an appropriate reference signal is not available. The filter is adapted in such a way as to restore some property of the signal lost before reaching the adaptive filter. Instead of the desired waveform as a template, as in the LMS or RLS algorithms, this property is used for the adaptation of the filter. When the desired signal is available, the conventional approach such as the LMS can be used; otherwise, a priori knowledge about the signal is used.

7.3 PROGRAMMING EXAMPLES FOR NOISE CANCELLATION AND SYSTEM IDENTIFICATION

The following programming examples illustrate adaptive filtering using the least mean squares (LMS) algorithm. It is instructive to read the first example even though it does not use the DSK, since it illustrates the steps in the adaptive process.

Example 7.1: Adaptive Filter Using C Code Compiled with Borland C/C++ (Adaptc)

This example applies the LMS algorithm using a C-coded program compiled with Borland C/C++. It illustrates the following steps for the adaptation process using the adaptive structure in Figure 7.1:

1. Obtain a new sample for each, the desired signal d and the reference input to the adaptive filter x, which represents a noise signal.
2. Calculate the adaptive FIR filter's output y, applying (7.1) as in Chapter 4 with an FIR filter. In the structure of Figure 7.1, the overall output is the same as the adaptive filter's output y.
3. Calculate the error signal applying (7.2).
4. Update/replace each coefficient or weight applying (7.3).
5. Update the input data samples for the next time n, with a data move scheme used in Chapter 4. Such a scheme moves the data instead of a pointer.
6. Repeat the entire adaptive process for the next output sample point.

Figure 7.6 shows a listing of the program adaptc.c, which implements the LMS algorithm for the adaptive filter structure in Figure 7.1. A desired signal is chosen as $2\cos(2n\pi f/F_s)$, and a reference noise input to the adaptive filter is chosen as $\sin(2n\pi f/F_s)$, where f is 1 kHz and $F_s = 8$ kHz. The adaptation rate, filter order, number of samples are 0.01, 22, and 40, respectively.

The overall output is the adaptive filter's output y, which adapts or converges to the desired cosine signal d.

The source file was compiled with Borland's C/C++ compiler. Execute this program. Figure 7.7 shows a plot of the adaptive filter's output (y_out) converging to the desired cosine signal. Change the adaptation or convergence rate β to 0.02 and verify a faster rate of adaptation.

```
//Adaptc.c Adaptation using LMS without TI's compiler

#include <stdio.h>
#include <math.h>
#define beta 0.01                              //convergence rate
#define N 21                                    //order of filter
#define NS 40                                   //number of samples
#define Fs 8000                                 //sampling frequency
#define pi 3.1415926
#define DESIRED 2*cos(2*pi*T*1000/Fs)    //desired signal
#define NOISE sin(2*pi*T*1000/Fs)        //noise signal

main()
{
 long I, T;
 double D, Y, E;
 double W[N+1] = {0.0};
 double X[N+1] = {0.0};
 FILE *desired, *Y_out, *error;
 desired = fopen ("DESIRED", "w++");    //file for desired samples
 Y_out = fopen ("Y_OUT", "w++");        //file for output samples
 error = fopen ("ERROR", "w++");        //file for error samples
 for (T = 0; T < NS; T++)               //start adaptive algorithm
  {
   X[0] = NOISE;                         //new noise sample
   D = DESIRED;                          //desired signal
   Y = 0;                                //filter'output set to zero
   for (I = 0; I <= N; I++)
    Y += (W[I] * X[I]);                  //calculate filter output
   E = D - Y                            //calculate error signal
   for (I = N; I >= 0; I--)
    {
     W[I] = W[I] + (beta*E*X[I]);       //update filter coefficients
     if (I != 0)
     X[I] = X[I-1];                      //update data sample
    }
   fprintf (desired, "\n%10g   %10f", (float) T/Fs, D);
   fprintf (Y_out, "\n%10g   %10f", (float) T/Fs, Y);
   fprintf (error, "\n%10g   %10f", (float) T/Fs, E);
  }
 fclose (desired);
 fclose (Y_out);
 fclose (error);
}
```

FIGURE 7.6. Adaptive filter program compiled with Borland C/C++ (adaptc.c).

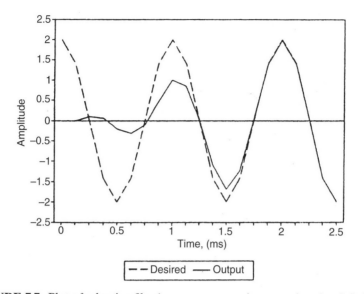

FIGURE 7.7. Plot of adaptive filter's output converging to cosine signal desired.

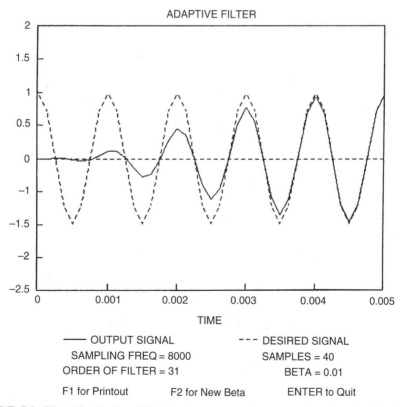

FIGURE 7.8. Plot of adaptive filter's output converging to cosine signal desired using interactive capability with progam `adaptive.c`.

Interactive Adaptation

A version of the program `adaptc.c` in Figure 7.6, with graphics and interactive capabilities to plot the adaptation process for different values of β is on the accompanying disk as `adaptive.c`, compiled with Turbo or Borland C/C++. It uses a desired cosine signal with an amplitude of 1 and a filter order of 31. Execute this program, enter a β value of 0.01, and verify the results in Figure 7.8. Note that the output converges to the desired cosine signal. Press F2 to execute this program again with a different beta value.

Example 7.2: Adaptive Filter for Noise Cancellation (`adaptnoise`)

This example illustrates the application of the LMS criterion to cancel an undesirable sinusoidal noise. Figure 7.9 shows a listing of the program `adaptnoise.c`, which implements an adaptive FIR filter using the structure in Figure 7.1. This program uses a float data format. An integer format version is included on the accompanying disk as `adaptnoise_int.c`.

A desired sine wave of 1500 Hz with an additive (undesired) sine wave noise of 312 Hz forms one of two inputs to the adaptive filter structure. A reference (template) cosine signal, with a frequency of 312 Hz, is the input to a 30-coefficient adaptive FIR filter. The 312-Hz reference cosine signal is correlated with the 312-Hz additive sine noise but not with the 1500-Hz desired sine signal.

For each time n, the output of the adaptive FIR filter is calculated and the 30 weights or coefficients are updated along with the delay samples. The "error" signal E is the overall desired output of the adaptive structure. This error signal is the difference between the desired signal and additive noise (`dplusn`), and the adaptive filter's output, `y(n)`.

All signals used are from a lookup table generated with MATLAB. No external inputs are used in this example. Figure 7.10 shows a MATLAB program `adaptnoise.m` (a more complete version is on the disk) that calculates the data values for the desired sine signal of 1500 Hz, the additive noise as a sine of 312 Hz, and the reference signal as a cosine of 312 Hz. The appropriate files generated (on the disk) are:

1. `dplusn`: sine(1500 Hz) + sine(312 Hz)
2. `refnoise`: cosine(312 Hz)

Figure 7.11 shows the file `sin1500.h` with sine data values that represent the 1500-Hz sine-wave signal desired. The frequency generated associated with `sin1500.h` is

$$f = F_s \,(\# \text{ of cycles})/(\# \text{ of points}) = 8000(24)/128 = 1500 \text{ Hz}$$

The constant `beta` determines the rate of convergence.

```
//Adaptnoise.c Adaptive FIR filter for noise cancellation

#include <refnoise.h>              //cosine 312 Hz
#include <dplusn.h>                //sin(1500) + sin(312)
#define beta 1E-9                  //rate of convergence
#define N 30                       //# of weights (coefficients)
#define NS 128                     //# of output sample points
float w[N];                        //buffer weights of adapt filter
float delay[N];                    //input buffer to adapt filter
short output;                      //overall output
short out_type = 1;                //output type for slider

interrupt void c_int11()           //ISR
{
 short i;
 static short buffercount=0;        //init count of # out samples
 float yn, E;                       //output filter/"error" signal

 delay[0] = refnoise[buffercount];  //cos(312 Hz) input to adapt FIR
 yn = 0;                            //init output of adapt filter

 for (i = 0; i < N; i++)            //to calculate out of adapt FIR
    yn += (w[i] * delay[i]);        //output of adaptive filter

 E = dplusn[buffercount] - yn;      //"error" signal=(d+n)-yn

 for (i = N-1; i >= 0; i--)         //to update weights and delays
   {
    w[i] = w[i] + beta*E*delay[i];  //update weights
    delay[i] = delay[i-1];          //update delay samples
   }
 buffercount++;                     //increment buffer count
 if (buffercount >= NS)             //if buffercount=# out samples
    buffercount = 0;                //reinit count

 if (out_type == 1)                 //if slider in position 1
    output = ((short)E*10);         //"error" signal overall output
 else if (out_type == 2)
    output=dplusn[buffercount]*10;  //desired(1500)+noise(312)

 output_sample(output);             //overall output result
 return;                            //return from ISR
}

void main()
{
 short T=0;
 for (T = 0; T < 30; T++)
   {
    w[T] = 0;                       //init buffer for weights
    delay[T] = 0;                   //init buffer for delay samples
   }
 comm_intr();                       //init DSK, codec, McBSP
 while(1);                          //infinite loop
}
```

FIGURE 7.9. Adaptive FIR filter program for noise cancellation (adaptnoise.c).

```
%Adaptnoise.m  Generates: dplusn.h, refnoise.h, sin1500.h

for i=1:128
  desired(i) = round(100*sin(2*pi*(i-1)*1500/8000));  %sin(1500)
  addnoise(i) = round(100*sin(2*pi*(i-1)*312/8000));  %sin(312)
  refnoise(i) = round(100*cos(2*pi*(i-1)*312/8000));  %cos(312)
end

dplusn = addnoise + desired;              %sin(312) + sin(1500)

fid=fopen('sin1500.h','w');               %desired sin(1500)
fprintf(fid,'short sin1500[128]={');
fprintf(fid,'%d,  ' ,desired(1:127));
fprintf(fid,'%d' ,desired(128));
fprintf(fid,'};\n');
fclose(fid);

% fid=fopen('dplusn.h','w');              %desired + noise
% fid=fopen('refnoise.h','w');            %reference noise
```

FIGURE 7.10. MATLAB program to generate data values for sine(1500), sine(1500) + sine(312), and cosine(312) (adaptnoise.m).

```
short sin1500[128]={0, 92, 71, -38, -100, -38, 71, 92, 0, -92, -71, 38,
100, 38, -71, -92, 0, 92, 71, -38, -100, -38, 71, 92, 0, -92, -71, 38,
100, 38, -71, -92, 0, 92, 71, -38, -100, -38, 71, 92, 0, -92, -71, 38,
100, 38, -71, -92, 0, 92, 71, -38, -100, -38, 71, 92, 0, -92, -71, 38,
100, 38, -71, -92, 0, 92, 71, -38, -100, -38, 71, 92, 0, -92, -71, 38,
100, 38, -71, -92, 0, 92, 71, -38, -100, -38, 71, 92, 0, -92, -71, 38,
100, 38, -71, -92, 0, 92, 71, -38, -100, -38, 71, 92, 0, -92, -71, 38,
100, 38, -71, -92, 0, 92, 71, -38, -100, -38, 71, 92, 0, -92, -71, 38,
100, 38, -71, -92};
```

FIGURE 7.11. MATLAB's header file generated for sine(1500 Hz) with 128 points (sin1500.h).

Build and run this project as **adaptnoise**. Verify the following output result: The undesired 312-Hz sinusoidal signal is being gradually reduced (canceled), while the desired 1500-Hz signal remains. Note that in this application the output desired is the error signal E, which adapts (converges) to the desired signal. A faster rate of cancellation can be observed with a larger value of *beta*. However, if *beta* is too large, the adaptation process will not be observed since the output would be shown as the 1500-Hz signal. With the slider is position 2, the output is (*dplusn*), the desired 1500-Hz sinusoidal signal with the additive 312-Hz noise signal.

Example 7.3: Adaptive FIR Filter for System ID of Fixed FIR (adaptIDFIR)

Figure 7.12 shows a listing of the program *adaptIDFIR.c*, which models or identifies an unknown system. See also Example 7.2, which implements an adaptive FIR for noise cancellation.

To test the adaptive scheme, the unknown system to be identified is chosen as an FIR bandpass filter with 55 coefficients centered at $F_s/4 = 2\,\text{kHz}$. The coefficients of this fixed FIR filter are in the file *bp55.cof*, introduced in Chapter 4. A 60-coefficient adaptive FIR filter models the fixed unknown FIR bandpass filter.

A pseudorandom noise sequence is generated within the program (see Examples 2.16 and 4.4) and becomes the input to both the fixed (unknown) and the adaptive FIR filters. This input signal represents a training signal. The adaptation process continues until the error signal is minimized. This feedback error signal is the difference between the output of the fixed unknown FIR filter and the output of the adaptive FIR filter.

An extra memory location is used in each of the two delay sample buffers (fixed and adaptive FIR). This is used to update the delay samples (see method B in Example 4.8).

Build and run this project as **adaptIDFIR** (using the C67x floating-point tools). Verify that the output (*adaptfir_out*) of the adaptive FIR filter is a bandpass filter centered at 2 kHz (with the slider in position 1 by default). With the slider in position 2, verify the output (*fir_out*) of the fixed FIR bandpass filter centered at 2 kHz and represented by the coefficient file *bp55.cof*. It can be observed that this output is practically identical to the adaptive filter's output.

Edit the main program to include the coefficient file *BS55.cof* (introduced in Example 4.4), which represents an FIR bandstop filter with 55 coefficients centered at 2 kHz. The FIR bandstop filter represents the unknown system to be identified.

Rebuild/run and verify that the output of the adaptive FIR filter (with the slider in position 1) is practically identical to the FIR bandstop filter (with the slider in position 2). Increase (decrease) *beta* by a factor of 10 to observe a faster (slower) rate of convergence. Change the number of weights (coefficients) from 60 to 40 and verify a slight degradation of the identification process.

Example 7.4: Adaptive FIR for System ID of Fixed FIR with Weights of Adaptive Filter Initialized as an FIR Bandpass (adaptIDFIRw)

The program *adaptIDFIR.c* in Example 7.3 is modified slightly to create the program *adaptIDFIRW.c* (on the accompanying disk). This new program initializes the weights of the adaptive FIR filter with the coefficients of an FIR bandpass filter centered at 3 kHz and represented by the coefficient file *bp3000.cof* (on the disk). The weights *w[i]* within the function *main* are initialized with the coefficients in the file *bp3000.cof* in lieu of zero.

```
//AdaptIDFIR.c Adaptive FIR for system ID of an FIR (uses C67 tools)

#include "bp55.cof"                    //fixed FIR filter coefficients
#include "noise_gen.h"                 //support noise generation file
#define beta 1E-13                     //rate of convergence
#define WLENGTH 60                     //# of coefffor adaptive FIR
float w[WLENGTH+1];                    //buffer coeff for adaptive FIR
int dly_adapt[WLENGTH+1];             //buffer samples of adaptive FIR
int dly_fix[N+1];                      //buffer samples of fixed FIR
short out_type = 1;                    //output for adaptive/fixed FIR
int fb;                                //feedback variable
shift_reg sreg;                        //shift register

int prand(void)                        //pseudo-random sequence {-1,1}
{
  int prnseq;
  if(sreg.bt.b0)
      prnseq = -8000;                  //scaled negative noise level
  else
      prnseq = 8000;                   //scaled positive noise level
  fb =(sreg.bt.b0)^(sreg.bt.b1);       //XOR bits 0,1
  fb^=(sreg.bt.b11)^(sreg.bt.b13);     //with bits 11,13 -> fb
  sreg.regval<<=1;
  sreg.bt.b0=fb;                       //close feedback path
  return prnseq;                       //return noise sequence
}

interrupt void c_int11()               //ISR
{
 int i;
 int fir_out = 0;                      //init output of fixed FIR
 int adaptfir_out = 0;                 //init output of adapt FIR
 float E;                              //error=diff of fixed/adapt out

 dly_fix[0] = prand();                 //input noise to fixed FIR
 dly_adapt[0]=dly_fix[0];              //as well as to adaptive FIR

 for (i = N-1; i>= 0; i--)
  {
   fir_out +=(h[i]*dly_fix[i]);        //fixed FIR filter output
   dly_fix[i+1] = dly_fix[i];          //update samples of fixed FIR
  }
```

FIGURE 7.12. Program to implement adaptive FIR filter that models (identifies) a fixed FIR filter (adaptIDFIR.c).

```
for (i = 0; i < WLENGTH; i++)
  adaptfir_out +=(w[i]*dly_adapt[i]);     //adaptive FIR filter output

E = fir_out - adaptfir_out;               //error signal

for (i = WLENGTH-1; i >= 0; i--)
  {
  w[i] = w[i]+(beta*E*dly_adapt[i]);      //update weights of adaptive FIR
  dly_adapt[i+1] = dly_adapt[i];          //update samples of adaptive FIR
  }

if (out_type == 1)                        //slider position for adapt FIR
  output_sample(adaptfir_out);            //output of adaptive FIR filter
else if (out_type == 2)                   //slider position for fixed FIR
  output_sample(fir_out);                 //output of fixed FIR filter
return;
}

void main()
{
int T=0, i=0;
for (i = 0; i < WLENGTH; i++)
  {
  w[i] = 0.0;                             //init coeff for adaptive FIR
  dly_adapt[i] = 0;                       //init buffer for adaptive FIR
  }
for (T = 0; T < N; T++)
  dly_fix[T] = 0;                         //init buffer for fixed FIR

sreg.regval=0xFFFF;                       //initial seed value
fb = 1;                                   //initial feevack value
comm_intr();                              //init DSK, codec, McBSP
while (1);                                //infinite loop
}
```

FIGURE 7.12. (*Continued*)

Build this project as **adaptIDFIRw** (using the C67x floating-point tools). Initially, the spectrum of the output of the adaptive FIR filter shows the FIR bandpass filter centered at 3 kHz. Then, gradually, the output spectrum adapts (converges) to the fixed (unknown) FIR bandpass filter centered at 2 kHz (represented by *bp55.cof*), while the reference filter gradually phases out. As the adaptation process takes place, one can observe at some time the two bandpass filters. You may wish to increase slightly the rate of adaptation (*beta*).

The adaptation process is illustrated with the CCS plots in Figure 7.13. Figure 7.14 illustrates the real-time adaptation process using an HP dynamic signal analyzer.

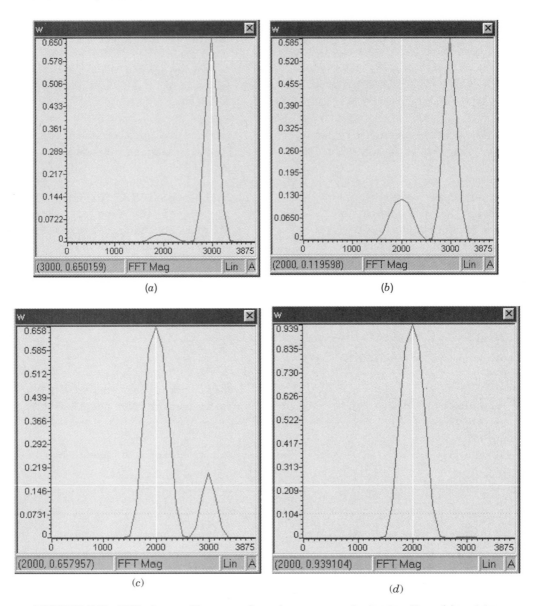

FIGURE 7.13. CCS plots to illustrate adaptation process of adaptive filter: (*a*) weights set initially as a 3-kHz bandpass filter; (*b*) weights starting to converge to a 2-kHz filter; (*c*) weights almost converged to 2kHz with the 3-kHz filter reduced; (*d*) adaptation completed with convergence to the 2-kHz bandpass filter.

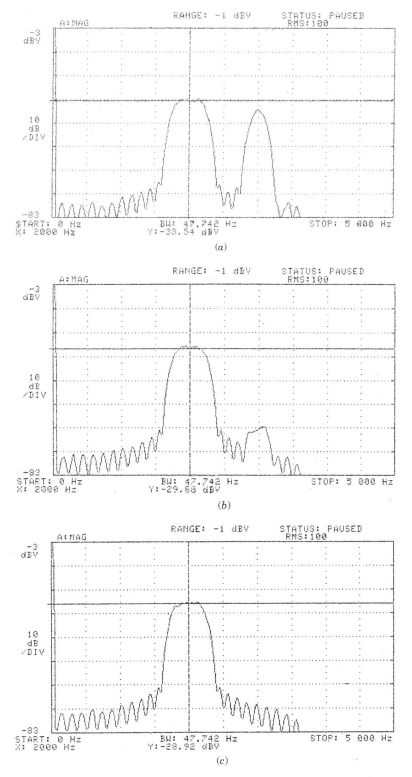

FIGURE 7.14. Real time adaptation process with adaptive filter converging to 2 kHz, obtained with an HP dynamic signal analyzer: (*a*) showing both the 3- and 2-kHz filters; (*b*) converging further to the 3-kHz filter; (*c*) adapted to the 3-kHz fixed filter.

Example 7.5: Adaptive FIR for System ID of Fixed IIR (adaptIDIIR)

Figure 7.15 shows a listing of the program *adaptIDIIR.c*, which uses an adaptive FIR filter to model or identify a system (fixed unknown IIR). See Example 5.1, which implements an IIR filter, and Examples 7.3 and 7.4, which implement an adaptive FIR filter to model a fixed FIR filter.

To test the adaptive scheme, the unknown system to be identified is chosen as a 36th-order IIR bandpass filter with *eighteen* second-order stages centered at 2 kHz. The coefficients of this fixed IIR filter are in the file *bp2000.cof*, introduced in Example 5.1. A 200-coefficient adaptive FIR filter is to model the fixed unknown IIR bandpass filter. A larger number of coefficients or weights than for the adaptive FIR filter are necessary for a good model of the IIR filter.

A pseudorandom noise sequence is generated (see Example 2.16) and becomes the input to both the fixed IIR filter and the adaptive FIR filter. The adaptation process continues until the error signal is minimized. This feedback error signal is the difference between the output of the fixed unknown IIR filter and the output of the adaptive FIR filter.

Build and run this project as **adaptIDIIR** (using the C67x floating-point tools). Verify that the output (*adaptfir_out*) converges to (models) the IIR bandpass filter centered at 2 kHz (with the slider initially in position 1). Verify that the output (*iir_out*) is the fixed IIR bandpass filter with the slider in position 2.

Include the coefficient file *lp2000.cof* in lieu of *bp2000.cof*. The coefficient file *lp2000.cof* represents an eighth-order (four second-order stages) IIR lowpass filter with a cutoff frequency of 2 kHz, introduced in Example 5.1. Verify that the adaptive FIR filter now adapts to the IIR lowpass filter with a cutoff frequency of 2 kHz.

Example 7.6: Adaptive Predictor for Cancellation of Narrowband Interference Added to Desired Wideband Signal (adaptpredict)

The program *adaptpredict.c*, shown in Figure 7.16, implements an adaptive FIR predictor for the cancellation of a narrowband interference in the presence of a wideband signal. The desired wideband signal with an additive narrowband interference is delayed and becomes the input to a 60-coefficient adaptive FIR filter.

The desired wideband signal is generated with a MATLAB program *wbsignal.m*, shown in Figure 7.17. This MATLAB program generates a 256-point lookup table in the file *wbsignal.h* (on the disk). A random sequence {-1,1} is generated, scaled, and written into the file *wbsignal.h*. Since the random sequence is for a length of 128 with a bit rate of 4 kHz, it is up-sampled to a 256-point sequence with a bit rate of 8 kHz. The wideband random sequence generated (with the file wbsignal.h) represents the signal desired.

The narrowband interference is an external signal. The bandwidth of the interference is narrow compared with the bandwidth of the random sequence generated

```
//AdaptIDIIR.c Adaptive FIR for system ID of fixed IIR using C67x tools

#include "bp2000.cof"              //BP @ 2kHz fixed IIR coeff
#include "noise_gen.h"             //support file noise sequence
#define beta 1E-11                 //rate of convergence
#define WLENGTH 200                //# of coeff for adaptive FIR
float w[WLENGTH+1];                //buffer coeff for adaptive FIR
int dly_adapt[WLENGTH+1];          //buffer samples of adaptive FIR
int dly_fix[stages][2] = {0};      //delay samples of fixed IIR
int a[stages][3], b[stages][2];    //coefficients of fixed IIR
short out_type = 1;                //slider adaptive FIR/fixed IIR
int fb;                            //feedback variable for noise
shift_reg sreg;                    //shift register for noise

int prand(void)                    //pseudo-random sequence {-1,1}
{
  int prnseq;
  if(sreg.bt.b0)
     prnseq = -4000;               //scaled negative noise level
  else
     prnseq= 4000;                 //scaled positive noise level
  fb =(sreg.bt.b0)^(sreg.bt.b1);   //XOR bits 0,1
  fb^=(sreg.bt.b11)^(sreg.bt.b13); //with bits 11,13 ->fb
  sreg.regval<<=1;
  sreg.bt.b0=fb;                   //close feedback path
  return prnseq;                   //return noise sequence
}

interrupt void c_int11()           //ISR
{
 int i, un, input, yn;
 int iir_out=0;                    //init output of fixed IIR
 int adaptfir_out=0;               //init output of adaptive FIR
 float E;                          //error signal

 dly_fix[0][0] = prand();          //input noise to fixed IIR
 dly_adapt[0] = dly_fix[0][0];     //same input to adaptive FIR
 input = prand();                  //noise as input to fixed IIR

 for (i = 0; i < stages; i++)      //repeat for each stage
  {
  un=input-((b[i][0]*dly_fix[i][0])>>15)-((b[i][1]*dly_fix[i][1])>>15);

  yn=((a[i][0]*un)>>15)+((a[i][1]*dly_fix[i][0])>>15)
     +((a[i][2]*dly_fix[i][1])>>15);
```

FIGURE 7.15. Program to implement adaptive FIR that models (identifies) a fixed IIR filter (adaptIDIIR.c).

```
dly_fix[i][1] = dly_fix[i][0];          //update delays of fixed IIR
dly_fix[i][0] = un;                     //update delays of fixed IIR
input = yn;                             //in next stage=out previous
}

iir_out = yn;                           //output of fixed IIR

for (i = 0; i < WLENGTH; i++)
  adaptfir_out +=(w[i]*dly_adapt[i]);   //output of adaptive FIR

E = iir_out - adaptfir_out;             //error as difference of outputs
for (i = WLENGTH; i > 0; i--)
 {
  w[i] = w[i]+(beta*E*dly_adapt[i]);    //update weights of adaptive FIR
  dly_adapt[i] = dly_adapt[i-1];        //update samples of adaptive FIR
 }

if (out_type == 1)                      //slider adaptive FIR/fixed IIR
  output_sample(adaptfir_out);          //output of adaptive FIR
else if (out_type == 2)
  output_sample(iir_out);               //output of fixed IIR
return;                                 //return to main
}

void main()
{
 int i=0;
 for (i = 0; i < WLENGTH; i++)
  {
  w[i] = 0.0;                           //init coeff of adaptive FIR
  dly_adapt[i] = 0.0;                   //init samples of adaptive FIR
  }
 sreg.regval=0xFFFF;                    //initial seed value
 fb = 1;                                //initial feedback value
 comm_intr();                           //init DSK, codec, McBSP
 while (1);                             //infinite loop
}
```

FIGURE 7.15. (*Continued*)

(the wideband signal desired). As a result, the samples of the interference are highly correlated. On the other hand, the samples of the wideband signal are relatively uncorrelated.

The characteristics of the narrowband interference permits the estimation of the narrowband interference from past samples of *splusn* in the program. The signal *splusn*, which represents the desired wideband signal with an additive narrowband

```
//Adaptpredict.C Adaptive predictor to cancel interference

#include "wbsignal.h"                  //wide-band signal table look-up
#define beta 1E-14                     //rate of convergence
#define N 60                           //# of coefficients of adapt FIR
const short bufferlength = NS;         //buffer length for wideband signal
short splusn[N+1];                     //buffer wideband signal+interference
float w[N+1];                          //buffer for weights of adapt FIR
float delay[N+1];                      //buffer for input to adapt FIR

interrupt void c_int11()               //ISR
{
 static short buffercount=0;           //init buffer
 int i;
 float yn, E;                          //yn=out adapt FIR, error signal
 short wb_signal;                      //wideband desired signal
 short noise;                          //external interference

 wb_signal=wbsignal[buffercount];      //wideband signal from look-up table
 noise = input_sample();               //external input as interference
 splusn[0] = wb_signal + noise;        //wideband signal+interference
 delay[0] = splusn[3];                 //delayed input to adaptive FIR
 yn = 0;                               //init output of adaptive FIR

 for (i = 0; i < N; i++)
   yn += (w[i] * delay[i]);            //output of adaptive FIR filter
 E = splusn[0] - yn;                   //(wideband+noise)-out adapt FIR

 for (i = N-1; i >= 0; i--)
  {
   w[i] = w[i]+(beta*E*delay[i]);      //update weights of adapt FIR
   delay[i+1] = delay[i];              //update buffer delay samples
   splusn[i+1] = splusn[i];            //update buffer corrupted wideband
  }

 buffercount++;                        //incr buffer count of wideband
 if (buffercount >= bufferlength)      //if buffer count=length of buffer
   buffercount = 0;                    //reinit count
 output_sample((short)E);             //overall output
 return;
}

void main()
{
 int T = 0;
 for (T = 0; T < N; T++)               //init variables
  {
   w[T] = 0.0;                         //buffer for weights of adaptive FIR
   delay[T] = 0.0;                     //buffer for delay samples
   splusn[T] = 0;                      //buffer for wideband+interference
  }
 comm_intr();                          //init DSK, codec, McBSP
 while(1);                             //infinite loop
}
```

FIGURE 7.16. Adaptive predictor program for cancellation of narrowband interference in the presence of a wideband signal (adaptpredict.c).

%wbsignal.m Generates wideband random sequence. Represents one info bit

```
len_code = 128;                        %length of random sequence
code = 2*round(rand(1,len_code))-1;    %generates random sequence {1,-1}
sample_rate = 2;                       %up-sampling from 4 to 8 kHz
NS = len_code * sample_rate;           %length of up-sampled sequence
sig = zeros(1,NS);                     %initialize random sequence
for i = 1:len_code                     %obtain up-sampled random sequence
   sig((i-1)*sample_rate + 1:i*sample_rate) = code(i);
end;
wbsignal = sig*5000;                   %scale for p-p amplitude of 500 mV

fid=fopen('wbsignal.h','w');           %open file for wideband signal
fprintf(fid,'#define NS 256 //number of output sample points\n\n');
fprintf(fid,'short wbsignal[256]={');
fprintf(fid,'%d, ',wbsignal(1:NS-1));
fprintf(fid,'%d',wbsignal(NS));
fprintf(fid,'};\n\n');
fclose(fid);
return;
```

FIGURE 7.17. MATLAB program to generate a desired wideband random sequence (wbsignal.m).

interference, is delayed before becoming the input to the adaptive FIR filter. The delay is sufficiently long so that the delayed wideband signal is uncorrelated with the undelayed sample.

The output of the adaptive FIR filter is an estimate of the correlated narrowband interference. As a result, the error signal E is an estimate of the wideband signal desired.

Build and run this project as **adaptpredict** (using the C67x floating-point tools). Apply a sinusoidal input signal between 1 and 3 kHz, representing the narrowband interference. Run the program and verify that the output spectrum of the error signal E adapts (converges) to the desired wideband signal, showing the input interference being gradually reduced.

Change the frequency of the input sinusoidal external interference and observe the adaptation process repeated to cancel the undesirable external interference. A faster rate of convergence can be observed by increasing beta by 10.

The wideband signal desired can be observed by outputting wb_signal (in lieu of E). Furthermore, the wideband signal with additive interference can be observed using output_sample($splusn[0]$). Better results are obtained when the amplitude of the external sinusoidal interference is about three times the amplitude of the wideband signal desired.

REFERENCES

1. B. Widrow and S. D. Stearns, *Adaptive Signal Processing*, Prentice Hall, Upper Saddle River, NJ, 1985.

2. B. Widrow and M. E. Hoff, Jr., Adaptive switching circuits, *IRE WESCON*, 1960, pp. 96–104.

3. B. Widrow, J. R. Glover, J. M. McCool, J. Kaunitz, C. S. Williams, R. H. Hearn, J. R. Zeidler, E. Dong, Jr., and R. C. Goodlin, Adaptive noise cancelling: principles and applications, *Proceedings of the IEEE*, Vol. 63, 1975, pp. 1692–1716.

4. R. Chassaing, *Digital Signal Processing with C and the TMS320C30*, Wiley, New York, 1992.

5. D. G. Manolakis, V. K. Ingle, and S. M. Kogon, *Statistical and Adaptive Signal Processing*, McGraw-Hill, New York, 2000.

6. S. Haykin, *Adaptive Filter Theory*, Prentice Hall, Upper Saddle River, NJ, 1986.

7. J. R. Treichler, C. R. Johnson, Jr., and M. G. Larimore, *Theory and Design of Adaptive Filters*, Wiley, New York, 1987.

8. S. M. Kuo and D. R. Morgan, *Active Noise Control Systems*, Wiley, New York, 1996.

9. K. Astrom and B. Wittenmark, *Adaptive Control*, Addison-Wesley, Reading, MA, 1995.

10. J. Tang, R. Chassaing, and W. J. Gomes III, Real-time adaptive PID controller using the TMS320C31 DSK, *Proceedings of the 2000 Texas Instruments DSPS Fest Conference*, 2000.

11. R. Chassaing, *Digital Signal Processing Laboratory Experiments Using C and the TMS320C31 DSK*, Wiley, New York, 1999.

12. R. Chassaing et al., Student projects on applications in digital signal processing with C and the TMS320C30, *Proceedings of the 2nd Annual TMS320 Educators Conference*, Texas Instruments, Dallas, TX, 1992.

13. C. S. Linquist, *Adaptive and Digital Signal Processing*, Steward and Sons, 1989.

14. S. D. Stearns and D. R. Hush, *Digital Signal Analysis*, Prentice Hall, Upper Saddle River, NJ, 1990.

15. J. R. Zeidler, Performance analysis of LMS adaptive prediction filters, *Proceedings of the IEEE*, Vol. 78, 1990, pp. 1781–1806.

16. S. T. Alexander, *Adaptive Signal Processing: Theory and Applications*, Springer-Verlag, New York, 1986.

17. C. F. Cowan and P. F. Grant, eds., *Adaptive Filters*, Prentice Hall, Upper Saddle River, NJ, 1985.

18. M. L. Honig and D. G. Messerschmitt, *Adaptive Filters: Structures, Algorithms and Applications*, Kluwer Academic, Norwell, MA, 1984.

19. V. Solo and X. Kong, *Adaptive Signal Processing Algorithms: Stability and Performance*, Prentice Hall, Upper Saddle River, NJ, 1995.

20. S. Kuo, G. Ranganathan, P. Gupta, and C. Chen, Design and implementation of adaptive filters, *IEEE 1988 International Conference on Circuits and Systems*, June 1988.

21. M. G. Bellanger, *Adaptive Digital Filters and Signal Analysis*, Marcel Dekker, New York, 1987.

22. R. Chassaing and B. Bitler, Adaptive filtering with C and the TMS320C30 digital signal processor, *Proceedings of the 1992 ASEE Annual Conference*, June 1992.

23. R. Chassaing, D. W. Horning, and P. Martin, Adaptive filtering with the TMS320C25, *Proceedings of the 1989 ASEE Annual Conference*, June 1989.

8

Code Optimization

- Optimization techniques for code efficiency
- Intrinsic C functions
- Parallel instructions
- Word-wide data access
- Software pipelining

In this chapter we illustrate several schemes that can be used to optimize and drastically reduce the execution time of your code. These techniques include the use of instructions in parallel, word-wide data, intrinsic functions, and software pipelining.

8.1 INTRODUCTION

Begin at a workstation level; for example, use C code on a PC. While code written in assembly (ASM) is processor-specific, C code can readily be ported from one platform to another. However, optimized ASM code runs faster than C and requires less memory space.

Before optimizing, make sure that the code is functional and yields correct results. After optimizing, the code can be so reorganized and resequenced that the optimization process makes it difficult to follow. One needs to realize that if a C-coded algorithm is functional and its execution speed is satisfactory, there is no need to optimize further.

After testing the functionality of your C code, transport it to the C6x platform. A floating-point implementation can be modeled first, then converted to a fixed-point implementation if desired. If the performance of the code is not adequate, use different compiler options to enable software pipelining (discussed later), reduce redundant loops, and so on. If the performance desired is still not achieved, you can use loop unrolling to avoid overhead in branching. This generally improves the execution speed but increases code size. You also can use word-wide optimization by loading/accessing 32-bit word (int) data rather than 16-bit half-word (short) data. You can then process lower and upper 16-bit data independently.

If performance is still not satisfactory, you can rewrite the time-critical section of the code in linear assembly, which can be optimized by the assembler optimizer. The profiler can be used to determine the specific function(s) that need to be optimized further.

The final optimization procedure that we discuss is a software pipelining scheme to produce hand-coded ASM instructions [1,2]. It is important to follow the procedure associated with software pipelining to obtain an efficient and optimized code.

8.2 OPTIMIZATION STEPS

If the performance and results of your code are satisfactory after any particular step, you are done.

1. Program in C. Build your project without optimization.
2. Use intrinsic functions when appropriate as well as the various optimization levels.
3. Use the profiler to determine/identify the function(s) that may need to be further optimized. Then convert these function(s) in linear ASM.
4. Optimize code in ASM.

8.2.1 Compiler Options

When the optimizer is invoked, the following steps are performed. A C-coded program is first passed through a parser that performs preprocessing functions and generates an intermediate file (.if) which becomes the input to an optimizer. The optimizer generates an .opt file which becomes the input to a code generator for further optimizations and generates an ASM file.

The options:

1. −o0 optimizes the use of registers.
2. −o1 performs a local optimization in addition to optimizations performed by the previous option: −o0.
3. −o2 performs a global optimization in addition to the optimizations performed by the previous options: −o0 and −o1.

4. -o3 performs a file optimization in addition to the optimizations performed by the three previous options: -o0, -o1, and -o2.

The options -o2 and -o3 attempt to do software optimization.

8.2.2 Intrinsic C Functions

There are a number of available C intrinsic functions that can be used to increase the efficiency of code (see also Example 3.1):

1. *int_mpy()* has the equivalent ASM instruction MPY, which multiplies the 16 LSBs of a number by the 16 LSBs of another number.
2. *int_mpyh()* has the equivalent ASM instruction MPYH, which multiplies the 16 MSBs of a number by the 16 MSBs of another number.
3. *int_mpylh()* has the equivalent ASM instruction MPYLH, which multiplies the 16 LSBs of a number by the 16 MSBs of another number.
4. *int_mpyhl()* has the equivalent instruction MPYHL, which multiplies the 16 MSBs of a number by the 16 LSBs of another number.
5. *void_nassert(int)* generates no code. It tells the compiler that the expression declared with the assert function is true. This conveys information to the compiler about alignment of pointers and arrays and of valid optimization schemes, such as word-wide optimization.
6. *uint_lo(double)* and *uint_hi(double)* obtain the low and high 32 bits of a double word, respectively (available on C67x or C64x).

8.3 PROCEDURE FOR CODE OPTIMIZATION

1. Use instructions in parallel so that multiple functional units can be operated within the same cycle.
2. Eliminate NOPs or delay slots, placing code where the NOPs are.
3. Unroll the loop to avoid overhead with branching.
4. Use word-wide data to access a 32-bit word (int) in lieu of a 16-bit half-word (short).
5. Use software pipelining, illustrated in Section 8.5.

8.4 PROGRAMMING EXAMPLES USING CODE OPTIMIZATION TECHNIQUES

Several examples are developed to illustrate various techniques to increase the efficiency of code. Optimization using software pipelining is discussed in Section 8.5.

The dot product is used to illustrate the various optimization schemes. The dot product of two arrays can be useful for many DSP algorithms, such as filtering and correlation. The examples that follow assume that each array consists of 200 numbers. Several programming examples using mixed C and ASM code, which provide necessary background, were given in Chapter 3.

Example 8.1: Sum of Products with Word-Wide Data Access for Fixed-Point Implementation Using C Code (twosum)

Figure 8.1 shows the C code *twosum.c*, which obtains the sum of products of two arrays accessing 32-bit word data. Each array consists of 200 numbers. Separate sums of products of even and odd terms are calculated within the loop. Outside the loop, the final summation of the even and odd terms is obtained.

For a floating-point implementation, the function and the variables *sum*, *suml*, and *sumh* in Figure 8.1 are cast as *float*, in lieu of *int*:

```
float   dotp (float a[ ], float b [ ])
{
      float suml, sumh, sum;
      int i;
      .
      .
      .
}
```

```
//twosum.c Sum of Products with separate accumulation of even/odd terms
//with word-wide data for fixed-point implementation

int     dotp (short a[ ], short b [ ])
{
      int suml, sumh, sum, i;
      suml = 0;
      sumh = 0;
      sum = 0;

      for (i = 0; i < 200; i +=2)
       {
        suml += a[i] * b[i];              //sum of products of even terms
        sumh += a[i + 1] * b[i + 1];      //sum of products of odd terms
       }
      sum = suml + sumh;                   //final sum of odd and even terms
      return (sum);
}
```

FIGURE 8.1. C code for sum of products using word-wide data access for separate accumulation of even and odd sum of products terms (twosum.c).

```
//dotpintrinsic.c  Sum of products with C intrinsic functions using C

for (i = 0; i < 100; i++)
        {
            sum1 = sum1 + _mpy(a[i], b[i]);
            sumh = sumh + _mpyh(a[i], b[i]);
        }
return (sum1 + sumh);
```

FIGURE 8.2. Separate sum of products using C intrinsic functions (dotpintrinsic.c).

Example 8.2: Separate Sum of Products with C Intrinsic Functions Using C Code (dotpintrinsic)

Figure 8.2 shows the C code dotpintrinsic.c to illustrate the separate sum of products using two C intrinsic functions, _mpy and _mpyh, which have the equivalent ASM instructions MPY and MPYH, respectively. Whereas the even and odd sum of products are calculated within the loop, the final summation is taken outside the loop and returned to the calling function.

Example 8.3: Sum of Products with Word-Wide Access for Fixed-Point Implementation Using Linear ASM Code (twosumlasmfix.sa)

Figure 8.3 shows the linear ASM code twosumlasmfix.sa, which obtains two separate sums of products for a fixed-point implementation using linear ASM code. It is not necessary to specify either the functional units or NOPs. Furthermore, symbolic names can be used for registers. The LDW instruction is used to load a 32-bit word-wide data value (which must be word-aligned in memory when using LDW). Lower and upper 16-bit products are calculated separately. The two ADD instructions accumulate separately the even and odd sum of products.

```
;twosumlasmfix.sa  Sum of Products. Separate accum of even/odd terms
;With word-wide data for fixed-point implementation using linear ASM

loop:       LDW         *aptr++, ai        ;32-bit word ai
            LDW         *bptr++, bi        ;32-bit word bi
            MPY         ai, bi, prodl      ;lower 16-bit product
            MPYH        ai, bi, prodh      ;higher 16-bit product
            ADD         prodl, suml, suml  ;accum even terms
            ADD         prodh, sumh, sumh  ;accum odd terms
            SUB         count, 1, count    ;decrement count
   [count]  B           loop               ;branch to loop
```

FIGURE 8.3. Separate sum of products using linear ASM code for fixed-point implementation (twosumlasmfix.sa).

```
;twosumlasmfloat.sa  Sum of products. Separate accum of even/odd terms
;Using double-word load LDDW for floating-point implementation

loop:          LDDW       *aptr++, ai1:ai0      ;64-bit word ai0 and ai1
               LDDW       *bptr++, bi1:bi0      ;64-bit word bi0 and bi1
               MPYSP      ai0, bi0, prodl       ;lower 32-bit product
               MPYSP      ai1, bi1, prodh       ;hiagher 32-bit product
               ADDSP      prodl, suml, suml     ;accum 32-bit even terms
               ADDSP      prodh, sumh, sumh     ;accum 32-bit odd terms
               SUB        count, 1, count       ;decrement count
      [count]  B          loop                  ;branch to loop
```

FIGURE 8.4. Separate sum of products with LDDW using linear ASM code for floating-point implementation (twosumlasmfloat.sa).

```
;dotpnp.asm  ASM Code with no-parallel instructions for fixed-point

               MVK     .S1    200, A1      ;count into A1
               ZERO    .L1    A7           ;init A7 for accum

LOOP           LDH     .D1    *A4++,A2     ;A2=16-bit data pointed by A4
               LDH     .D1    *A8++,A3     ;A3=16-bit data pointed by A8
               NOP            4            ;4 delay slots for LDH
               MPY     .M1    A2,A3,A6     ;product in A6
               NOP                         ;1 delay slot for MPY
               ADD     .L1    A6,A7,A7     ;accum in A7
               SUB     .S1    A1,1,A1      ;decrement count
      [A1]     B       .S2    LOOP         ;branch to LOOP
               NOP            5            ;5 delay slots for B
```

FIGURE 8.5. ASM code with no parallel instructions for fixed-point implementation (dotpnp.asm).

Example 8.4: Sum of Products with Double-Word Load for Floating-Point Implementation Using Linear ASM Code (twosumlasmfloat)

Figure 8.4 shows the linear ASM code *twosumlasmfloat.sa* to obtain two separate sums of products for a floating-point implementation using linear ASM code. The double-word load instruction LDDW loads a 64-bit data value and stores it in a pair of registers. Each single-precision multiply instruction MPYSP performs a 32×32 multiplication. The sums of products of the lower and upper 32 bits are performed to yield a sum of both even and odd terms as 32 bits.

Example 8.5: Dot Product with No Parallel Instructions for Fixed-Point Implementation Using ASM Code (dotpnp)

Figure 8.5 shows the ASM code *dotpnp.asm* for the dot product with no instructions in parallel for a fixed-point implementation. A fixed-point implementation can

```
;dotpp.asm  ASM Code with parallel instructions for fixed-point

          MVK      .S1     200, A1        ;count into A1
  ||      ZERO     .L1     A7             ;init A7 for accum

LOOP      LDH      .D1     *A4++,A2       ;A2=16-bit data pointed by A4
  ||      LDH      .D2     *B4++,B2       ;B2=16-bit data pointed by B4
          SUB      .S1     A1,1,A1        ;decrement count
  [A1]    B        .S1     LOOP           ;branch to LOOP (after ADD)
          NOP              2              ;delay slots for LDH and B
          MPY      .M1x    A2,B2,A6       ;product in A6
          NOP                             ;1 delay slot for MPY
          ADD      .L1     A6,A7,A7       ;accum in A7,then branch
;branch occurs here
```

FIGURE 8.6. ASM code with parallel instructions for fixed-point implementation (dotpp.asm).

be performed with all C6x devices, whereas a floating-point implementation requires a C67x platform such as the C6711 DSK.

The loop iterates 200 times. With a fixed-point implementation, each pointer register A4 and A8 increments to point at the next half-word (16 bits) in each buffer, whereas with a floating-point implementation, a pointer register increments the pointer to the next 32-bit word. The load, multiply, and branch instructions must use the .D, .M, and .S units, respectively; the add and subtract instructions can use any unit (except .M). The instructions within the loop consume 16 cycles per iteration. This yields $16 \times 200 = 3200$ cycles. Table 8.4 shows a summary of several optimization schemes for both fixed- and floating-point implementations.

Example 8.6: Dot Product with Parallel Instructions for Fixed-Point Implementation Using ASM Code (dotpp)

Figure 8.6 shows the ASM code *dotpp.asm* for the dot product with a fixed-point implementation with instructions in parallel. With code in lieu of NOPs, the number of NOPs is reduced.

The MPY instruction uses a cross-path (with .M1x) since the two operands are from different register files or different paths. The instructions SUB and B are moved up to fill some of the delay slots required by LDH. The branch instruction occurs after the ADD instruction. Using parallel instructions, the instructions within the loop now consume eight cycles per iteration, to yield $8 \times 200 = 1600$ cycles.

Example 8.7: Two Sums of Products with Word-Wide (32-bit) Data for Fixed-Point Implementation Using ASM Code (twosumfix)

Figure 8.7 shows the ASM code *twosumfix.asm*, which calculates two separate sums of products using word-wide access of data for a fixed-point implementation. The loop count is initialized to 100 (not 200) since two sums of products are obtained

```
;twosumfix.asm  ASM code for two sums of products with word-wide data
;for fixed-point implementation

                MVK      .S1   100,A1      ;count/2 into A1
       ||       ZERO     .L1   A7          ;init A7 for accum of even terms
       ||       ZERO     .L2   B7          ;init B7 for accum of odd terms

LOOP            LDW      .D1   *A4++,A2    ;A2=32-bit data pointed by A4
       ||       LDW      .D2   *B4++,B2    ;A3=32-bit data pointed by B4
                SUB      .S1   A1,1,A1     ;decrement count
          [A1]  B        .S1   LOOP        ;branch to LOOP (after ADD)
                NOP            2           ;delay slots for both LDW and B
                MPY      .M1x  A2,B2,A6    ;lower 16-bit product in A6
       ||       MPYH     .M2x  A2,B2,B6    ;upper 16-bit product in B6
                NOP                        ;1 delay slot for MPY/MPYH
                ADD      .L1   A6,A7,A7    ;accum even terms in A7
       ||       ADD      .L2   B6,B7,B7    ;accum odd terms in B7
;branch occurs here
```

FIGURE 8.7. ASM code for two sums of products with 32-bit data for fixed-point implementation (twosumfix.asm).

per iteration. The instruction LDW loads a word or 32-bit data. The multiply instruction MPY finds the product of the lower 16×16 data, and MPYH finds the product of the upper 16×16 data. The two ADD instructions accumulate separately the even and odd sums of products. Note that an additional ADD instruction is needed outside the loop to accumulate A7 and B7. The instructions within the loop consume eight cycles, now using 100 iterations (not 200), to yield $8 \times 100 = 800$ cycles.

Example 8.8: Dot Product with No Parallel Instructions for Floating-Point Implementation Using ASM Code (dotpnpfloat)

Figure 8.8 shows the ASM code dotpnpfloat.asm for the dot product with a floating-point implementation using no instructions in parallel. The loop iterates 200 times. The single-precision floating-point instruction MPYSP performs a 32×32 multiply. Each MPYSP and ADDSP requires three delay slots. The instructions within the loop consume a total of 18 cycles per iteration (without including three NOPs associated with ADDSP). This yields a total of $18 \times 200 = 3600$ cycles. (See Table 8.4 for a summary of several optimization schemes for both fixed- and floating-point implementations.)

Example 8.9: Dot Product with Parallel Instructions for Floating-Point Implementation Using ASM Code (dotppfloat)

Figure 8.9 shows the ASM code dotppfloat.asm for the dot product with a floating-point implementation using instructions in parallel. The loop iterates 200

```
;dotpnpfloat.asm  ASM with no parallel instructions for floating-point

                MVK        .S1     200, A1         ;count into A1
                ZERO       .L1     A7              ;init A7 for accum

LOOP            LDW        .D1     *A4++,A2        ;A2=32-bit data pointed by A4
                LDW        .D1     *A8++,A3        ;A3=32-bit data pointed by A8
                NOP                4               ;4 delay slots for LDW
                MPYSP      .M1     A2,A3,A6        ;product in A6
                NOP                3               ;3 delay slots for MPYSP
                ADDSP      .L1     A6,A7,A7        ;accum in A7
                SUB        .S1     A1,1,A1         ;decrement count
        [A1]    B          .S2     LOOP            ;branch to LOOP
                NOP                5               ;5 delay slots for B
```

FIGURE 8.8. ASM code with no parallel instructions for floating-point implementation (dotpnpfloat.asm).

```
;dotppfloat.asm  ASM Code with parallel instructions for floating-point

                MVK        .S1     200, A1         ;count into A1
   ||           ZERO       .L1     A7              ;init A7 for accum

LOOP            LDW        .D1     *A4++,A2        ;A2=32-bit data pointed by A4
   ||           LDW        .D2     *B4++,B2        ;B2=32-bit data pointed by B4
                SUB        .S1     A1,1,A1         ;decrement count
                NOP                2               ;delay slots for both LDW and B
        [A1]    B          .S2     LOOP            ;branch to LOOP (after ADDSP)
                MPYSP      .M1x    A2,B2,A6        ;product in A6
                NOP                3               ;3 delay slots for MPYSP
                ADDSP      .L1     A6,A7,A7        ;accum in A7,then branch
;branch occurs here
```

FIGURE 8.9. ASM code with parallel instructions for floating-point implementation (dotppfloat.asm).

times. By moving the SUB and B instructions up to take the place of some NOPs, the number of instructions within the loop is reduced to 10. Note that three additional NOPs would be needed outside the loop to retrieve the result from ADDSP. The instructions within the loop consume a total of 10 cycles per iteration. This yields a total of $10 \times 200 = 2000$ cycles.

Example 8.10: Two Sums of Products with Double-Word-Wide (64-bit) Data for Floating-Point Implementation Using ASM Code (twosumfloat)

Figure 8.10 shows the ASM code twosumfloat.asm, which calculates two separate sums of products using double-word-wide access of 64-bit data for a floating-point implementation. The loop count is initialized to 100 since two sums of products are

```
;twosumfloat.asm  ASM Code for two sums of products for floating-point

              MVK       .S1    100,A1      ;count/2 into A1
       ||     ZERO      .L1    A7          ;init A7 for accum of even terms
       ||     ZERO      .L2    B7          ;init B7 for accum of odd terms

LOOP          LDDW      .D1    *A4++,A3:A2 ;64-bit into register pair A2,A3
       ||     LDDW      .D2    *B4++,B3:B2 ;64-bit into register pair B2,B3
              SUB       .S1    A1,1,A1     ;decrement count
              NOP              2           ;delay slots for LDW
      [A1]    B         .S2    LOOP        ;branch to LOOP
              MPYSP     .M1x   A2,B2,A6    ;lower 32-bit product in A6
       ||     MPYSP     .M2x   A3,B3,B6    ;upper 32-bit product in B6
              NOP              3           ;3 delay slot for MPYSP
              ADDSP     .L1    A6,A7,A7    ;accum even terms in A7
       ||     ADDSP     .L2    B6,B7,B7    ;accum odd terms in B7
;branch occurs here
              NOP              3           ;delay slots for last ADDSP
              ADDSP     .L1x   A7,B7,A4    ;final sum of even and odd terms
              NOP              3           ;delay slots for ADDSP
```

FIGURE 8.10. ASM code with two sums of products for floating-point implementation
(twosumfloat.asm).

obtained per iteration. The instruction LDDW loads a 64-bit double-word data value
into a register pair. The multiply instruction MPYSP performs a 32×32 multiply. The
two ADDSP instructions accumulate separately the even and odd sums of products.
The additional ADDSP instruction is needed outside the loop to accumulate A7 and
B7. The instructions within the loop consume a total of 10 cycles, using 100 iterations (not 200), to yield a total of $10 \times 100 = 1000$ cycles.

8.5 SOFTWARE PIPELINING FOR CODE OPTIMIZATION

Software pipelining is a scheme to write efficient code in ASM so that all the functional units are utilized within one cycle. Optimization levels $-o2$ and $-o3$ enable
code generation to generate (or attempt to generate) software-pipelined code.

There are three stages associated with software pipelining:

1. *Prolog (warm-up)*. This stage contains instructions needed to build up the
   loop kernel (cycle).
2. *Loop kernel (cycle)*. Within this loop, all instructions are executed in parallel.
   The entire loop kernel is executed in *one* cycle, since all the instructions within
   the loop kernel stage are in parallel.
3. *Epilog (cool-off)*. This stage contains the instructions necessary to complete
   all iterations.

8.5.1 Procedure for Hand-Coded Software Pipelining

1. Draw a dependency graph.
2. Set up a scheduling table.
3. Obtain code from the scheduling table.

8.5.2 Dependency Graph

Figure 8.11 shows a dependency graph. A procedure for drawing a dependency graph follows.

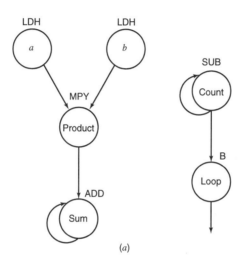

(a)

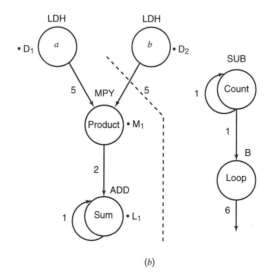

(b)

FIGURE 8.11. Dependency graph for dotp product: (a) initial stage; (b) final stage.

1. Draw the nodes and paths.
2. Write the number of cycles to complete an instruction.
3. Assign functional units associated with each node.
4. Separate the data path so that the maximum number of units are utilized.

A node has one or more data paths going in and/or out of the node. The numbers next to each node represent the number of cycles required to complete the associated instruction. A parent node contains an instruction that writes to a variable; whereas a child node contains an instruction that reads a variable written by the parent.

The LDH instructions are considered to be the parents of the MPY instruction since the results of the two load instructions are used to perform the MPY instruction. Similarly, the MPY is the parent of the ADD instruction. The ADD instruction is fed back as input for the next iteration; similarly with the SUB instruction.

Figure 8.12 shows another dependency graph associated with two sums of products for a fixed-point implementation. The length of the prolog section is the longest path from the dependency graph in Figure 8.12. Since the longest path is 8, the length of the prolog is 7, before entering the loop kernel (cycle) at cycle 8.

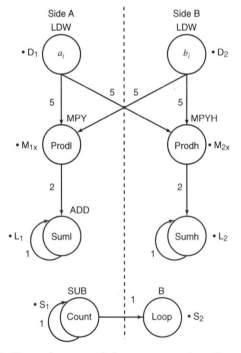

FIGURE 8.12. Dependency graph for two sums of products per iteration.

A similar dependency graph for a floating-point implementation can be obtained using LDW, MPYSP, and ADDSP in lieu of LDH, MPY, and ADD, respectively, in Figure 8.12. Note that the single-precision instructions ADDSP and MPYSP both take four cycles to complete (three delay slots each).

8.5.3 Scheduling Table

Table 8.1 shows a scheduling table drawn from the dependency graph.

1. LDW starts in cycle 1.
2. MPY and MPYH must start five cycles after the LDWs, due to the four delay slots. Therefore, MPY and MPYH start in cycle 6.
3. ADD must start two cycles after MPY/MPYH, due to the one delay slot of MPY/MPYH. Therefore, ADD starts in cycle 8.
4. B has five delay slots and starts in cycle 3, since branching occurs in cycle 9, after the ADD instruction.
5. SUB instruction must start one cycle before the branch instruction, since the loop count is decremented before branching occurs. Therefore, SUB starts in cycle 2.

From Table 8.1, the two LDW instructions are in parallel and are issued in cycles 1, 9, 17, ... The SUB instruction is issued in cycles 2, 10, 18, ... This is followed by the branch (B) instruction issued in cycles 3, 11, 19, ... The two parallel instructions MPY and MPYH are issued in cycles 6, 14, 22, ... The ADD instructions are issued in cycles 8, 16, 24, ...

Table 8.1 is extended to illustrate the different stages: prolog (cycles 1 through 7), loop kernel (cycle 8), and epilog (cycles 9, 10, . . . not shown), as shown in Table 8.2. The instructions within the prolog stage are repeated until and including the loop kernel (cycle) stage. Instructions in the epilog stage (cycles 9, 10, . . .) are to complete the functionality of the code.

From Table 8.2, an efficient optimized code can be obtained. Note that it is possible to start processing a new iteration before previous iterations are finished. Software pipelining allows us to determine when to start a new loop iteration.

Loop Kernel (Cycle)
Within the loop kernel, in cycle 8, each functional unit is used only once. The minimum iteration interval is the minimum number of cycles required to wait before the initiation of a successive iteration. This interval is 1. As a result, a new iteration can be initiated every cycle.

Within the loop cycle 8, multiple iterations of the loop execute in parallel. In

TABLE 8.1 Schedule Table of Dot Product before Software Pipelining for Fixed-Point Implementation

Units	Cycles							
	1, 9, ...	2, 10, ...	3, 11, ...	4, 12, ...	5, 13, ...	6, 14, ...	7, 15, ...	8, 16, ...
.D1	LDW							
.D2	LDW							
.M1						MPY		
.M2						MPYH		
.L1								ADD
.L2								ADD
.S1		SUB						
.S2			B					

TABLE 8.2 Schedule Table of Dot Product after Software Pipelining for Fixed-Point Implementation

Units	Cycles							
	Prolog							Loop Kernel
	1	2	3	4	5	6	7	8
.D1	LDW	LDW	LDW	LDW	LDW	LDW	LDW	LDW
.D2	LDW	LDW	LDW	LDW	LDW	LDW	LDW	LDW
.M1						MPY	MPY	MPY
.M2						MPYH	MPYH	MPYH
.L1								ADD
.L2								ADD
.S1		SUB	SUB	SUB	SUB	SUB	SUB	SUB
.S2			B	B	B	B	B	B

cycle 8, different iterations are processed at the same time. For example, the ADDs add data for iteration 1, while MPY and MPYH multiply data for iteration 3, LDWs load data for iteration 8, SUB decrements the counter for iteration 7, and B branches for iteration 6. Note that the values being multiplied are loaded into registers five cycles prior to the cycle when the values are multiplied. Before the first multiplication occurs, the fifth load has just completed. This software pipeline is eight iterations deep.

Example 8.11: Dot Product Using Software Pipelining for a Fixed-Point Implementation

This example implements the dot product using software pipelining for a fixed-point implementation. From Table 8.2, one can readily obtained the ASM code *dotpiped-*

fix.asm shown in Figure 8.13. The loop count is 100 since two multiplies and two accumulates are calculated per iteration. The following instructions start in the following cycles:

> *Cycle 1*: LDW, LDW (also initialization of count, and the accumulators A7 and B7)
> *Cycle 2*: LDW, LDW, SUB
> *Cycles 3–5*: LDW, LDW, SUB, B
> *Cycles 6–7*: LDW, LDW, MPY, MPYH, SUB, B
> *Cycles 8–107*: LDW, LDW, MPY, MPYH, ADD, ADD, SUB, B
> *Cycle 108*: LDW, LDW, MPY, MPYH, ADD, ADD, SUB, B

The prolog section is within cycles 1 through 7; the loop kernel is in cycle 8, where all the instructions are in parallel; and the epilog section is in cycle 108. Note that SUB is made conditional to ensure that A1 is no longer decremented once it reaches zero.

Example 8.12: Dot Product Using Software Pipelining for a Floating-Point Implementation

This example implements the dot product using software pipelining for a floating-point implementation. Table 8.3 shows a floating-point version of Table 8.2. LDW becomes LDDW, MPY/MPYH become MPYSP, and ADD becomes ADDSP. Both MPYSP and ADDSP have three delays slots. As a result, the loop kernel starts in cycle 10 (not cycle 8). The SUB and B instructions start in cycles 4 and 5, respectively, in lieu of cycles 2 and 3. ADDSP starts in cycle 10 in lieu of cycle 8. The software pipeline for a floating-point implementation is 10 deep.

TABLE 8.3   Schedule Table of Dot Product after Software Pipelining for Floating-Point Implementation

| | Cycle | | | | | | | | | |
| | Prolog | | | | | | | | | Loop Kernel |
Units	1	2	3	4	5	6	7	8	9	10
.D1	LDDW	LDDW	LDDW	LDDW	LDDW	LDDW	LDDW	LDDW	LDDW	LDDW
.D2	LDDW	LDDW	LDDW	LDDW	LDDW	LDDW	LDDW	LDDW	LDDW	LDDW
.M1						MPYSP	MPYSP	MPYSP	MPYSP	MPYSP
.M2						MPYSP	MPYSP	MPYSP	MPYSP	MPYSP
.L1										ADDSP
.L2										ADDSP
.S1				SUB	SUB	SUB	SUB	SUB	SUB	SUB
.S2					B	B	B	B	B	B

```
;dotpipedfix.asm ASM code for dot product with software pipelining
;For fixed-point implementation
;cycle 1
                MVK       .S1    100,A1          ;loop count
        ||      ZERO      .L1    A7              ;init accum A7
        ||      ZERO      .L2    B7              ;init accum B7
        ||      LDW       .D1    *A4++,A2        ;32-bit data in A2
        ||      LDW       .D2    *B4++,B2        ;32-bit data in B2
;cycle 2
        ||      LDW       .D1    *A4++,A2        ;32-bit data in A2
        ||      LDW       .D2    *B4++,B2        ;32-bit data in B2
        || [A1] SUB       .S1    A1,1,A1         ;decrement count
;cycle 3
        ||      LDW       .D1    *A4++,A2        ;32-bit data in A2
        ||      LDW       .D2    *B4++,B2        ;32-bit data in B2
        || [A1] SUB       .S1    A1,1,A1         ;decrement count
        || [A1] B         .S2    LOOP            ;branch to LOOP
;cycle 4
        ||      LDW       .D1    *A4++,A2        ;32-bit data in A2
        ||      LDW       .D2    *B4++,B2        ;32-bit data in B2
        || [A1] SUB       .S1    A1,1,A1         ;decrement count
        || [A1] B         .S2    LOOP            ;branch to LOOP
;cycle 5
        ||      LDW       .D1    *A4++,A2        ;32-bit data in A2
        ||      LDW       .D2    *B4++,B2        ;32-bit data in B2
        || [A1] SUB       .S1    A1,1,A1         ;decrement count
        || [A1] B         .S2    LOOP            ;branch to LOOP
;cycle 6
        ||      LDW       .D1    *A4++,A2        ;32-bit data in A2
        ||      LDW       .D2    *B4++,B2        ;32-bit data in B2
        || [A1] SUB       .S1    A1,1,A1         ;decrement count
        || [A1] B         .S2    LOOP            ;branch to LOOP
        ||      MPY       .M1x   A2,B2,A6        ;lower 16-bit product into A6
        ||      MPYH      .M2x   B2,A2,B6        ;upper 16-bit product into B6
;cycle 7
        ||      LDW       .D1    *A4++,A2        ;32-bit data in A2
        ||      LDW       .D2    *B4++,B2        ;32-bit data in B2
        || [A1] SUB       .S1    A1,1,A1         ;decrement count
        || [A1] B         .S2    LOOP            ;branch to LOOP
        ||      MPY       .M1x   A2,B2,A6        ;lower 16-bit product into A6
        ||      MPYH      .M2x   B2,A2,B6        ;upper 16-bit product into B6
;cycles 8-107 (loop cycle)
        ||      LDW       .D1    *A4++,A2        ;32-bit data in A2
        ||      LDW       .D2    *B4++,B2        ;32-bit data in B2
        || [A1] SUB       .S1    A1,1,A1         ;decrement count
        || [A1] B         .S2    LOOP            ;branch to LOOP
        ||      MPY       .M1x   A2,B2,A6        ;lower 16-bit product into A6
        ||      MPYH      .M2x   B2,A2,B6        ;upper 16-bit product into B6
        ||      ADD       .L1    A6,A7,A7        ;accum in A7
        ||      ADD       .L2    B6,B7,B7        ;accum in B7
;branch occurs here
;cycle 108 (epilog)
                ADD       .L1x   A7,B7,A4        ;final accum of odd/even
```

FIGURE 8.13. ASM code using software pipelining for fixed-point implementation (dot-pipedfix.asm).

Figure 8.14 shows the ASM code *dotpipedfloat.asm*, which implements the floating-point version of the dot product. Since ADDSP has three delay slots, the accumulation is staggered by four. The accumulation associated with one of the ADDSP instructions at each loop cycle follows:

Loop Cycle	Accumulator (one ADDSP)	
1	0	
2	0	
3	0	
4	0	
5	p0	;first product
6	p1	;second product
7	p3	
8	p4	
9	p0 + p4	;sum of first and fifth products
10	p1 + p5	;sum of second and sixth products
11	p2 + p6	
12	p3 + p7	
13	p0 + p4 + p8	;sum of first, fifth, and ninth products
14	p1 + p5 + p9	
15	p2 + p6 + p10	
16	p3 + p7 + p11	
17	p0 + p4 + p8 + p12	
⋮	⋮	
99	p2 + p6 + p10 + ... + p94	
100	p3 + p7 + p11 + ... + p95	

This accumulation is shown associated with the loop cycle. The actual cycle is shifted by 9 (by the cycles in the prolog section). Note that the first product, p0, is obtained (available) in loop cycle 5 since the first ADDSP starts in loop cycle 1 and has three delay slots. The first product, p0, is associated with the lower 32-bit term. The second ADDSP (not shown) accumulates the upper 32-bit sum of products.

A6 contains the lower 32-bit products and B6 contains the upper 32-bit products. The sum of the lower and upper 32-bit products are accumulated in A7 and B7, respectively.

The epilog section contains the following instructions associated with the actual cycle (not loop cycles), as shown in Figure 8.14.

```
;dotpipedfloat.asm  ASM code for dot product with software pipelining
;For floating-point implementation
;cycle 1
                MVK       .S1     100,A1        ;loop count
        ||      ZERO      .L1     A7            ;init accum A7
        ||      ZERO      .L2     B7            ;init accum B7
        ||      LDDW      .D1     *A4++,A3:A2   ;64-bit data in A2 and A3
        ||      LDDW      .D2     *B4++,B3:B2   ;64-bit data in B2 and B3
;cycle 2
        ||      LDDW      .D1     *A4++,A3:A2   ;64-bit data in A2 and A3
        ||      LDDW      .D2     *B4++,B3:B2   ;64-bit data in B2 and B3
;cycle 3
        ||      LDDW      .D1     *A4++,A3:A2   ;64-bit data in A2 and A3
        ||      LDDW      .D2     *B4++,B3:B2   ;64-bit data in B2 and B3
;cycle 4
        ||      LDDW      .D1     *A4++,A3:A2   ;64-bit data in A2 and A3
        ||      LDDW      .D2     *B4++,B3:B2   ;64-bit data in B2 and B3
        || [A1] SUB       .S1     A1,1,A1       ;decrement count
;cycle 5
        ||      LDDW      .D1     *A4++,A3:A2   ;64-bit data in A2 and A3
        ||      LDDW      .D2     *B4++,B3:B2   ;64-bit data in B2 and B3
        || [A1] SUB       .S1     A1,1,A1       ;decrement count
        || [A1] B         .S2     LOOP          ;branch to LOOP
;cycle 6
        ||      LDDW      .D1     *A4++,A3:A2   ;64-bit data in A2 and A3
        ||      LDDW      .D2     *B4++,B3:B2   ;64-bit data in B2 and B3
        || [A1] SUB       .S1     A1,1,A1       ;decrement count
        || [A1] B         .S2     LOOP          ;branch to LOOP
        ||      MPYSP     .M1x    A2,B2,A6      ;lower 32-bit product into A6
        ||      MPYSP     .M2x    B3,A3,B6      ;upper 32-bit product into B6
;cycle 7
        ||      LDDW      .D1     *A4++,A3:A2   ;32-bit data in A2 and A3
        ||      LDDW      .D2     *B4++,B3:B2   ;32-bit data in B2 and B3
        || [A1] SUB       .S1     A1,1,A1       ;decrement count
        || [A1] B         .S2     LOOP          ;branch to LOOP
        ||      MPYSP     .M1x    A2,B2,A6      ;lower 32-bit product into A6
        ||      MPYSP     .M2x    B3,A3,B6      ;upper 32-bit product into B6
;cycle 8
        ||      LDDW      .D1     *A4++,A3:A2   ;32-bit data in A2 and A3
        ||      LDDW      .D2     *B4++,B3:B2   ;32-bit data in B2 and B3
        || [A1] SUB       .S1     A1,1,A1       ;decrement count
        || [A1] B         .S2     LOOP          ;branch to LOOP
        ||      MPYSP     .M1x    A2,B2,A6      ;lower 32-bit product into A6
        ||      MPYSP     .M2x    B3,A3,B6      ;upper 32-bit product into B6
;cycle 9
        ||      LDDW      .D1     *A4++,A3:A2   ;32-bit data in A2 and A3
        ||      LDDW      .D2     *B4++,B3:B2   ;32-bit data in B2 and B3
        || [A1] SUB       .S1     A1,1,A1       ;decrement count
        || [A1] B         .S2     LOOP          ;branch to LOOP
        ||      MPYSP     .M1x    A2,B2,A6      ;lower 32-bit product into A6
        ||      MPYSP     .M2x    B3,A3,B6      ;upper 32-bit product into B6
```

FIGURE 8.14. ASM code using software pipelining for floating-point implementation (dotpipedfloat.asm).

```
;cycles 10-109 (loop kernel)
        ||        LDDW      .D1    *A4++,A3:A2   ;32-bit data in A2 and A3
        ||        LDDW      .D2    *B4++,B3:B2   ;32-bit data in B2 and B3
        || [A1]   SUB       .S1    A1,1,A1       ;decrement count
        || [A1]   B         .S2    LOOP          ;branch to LOOP
        ||        MPYSP     .M1x   A2,B2,A6      ;lower 32-bit product into A6
        ||        MPYSP     .M2x   B3,A3,B6      ;upper 32-bit product into B6
        ||        ADDSP     .L1    A6,A7,A7      ;accum in A7
        ||        ADDSP     .L2    B6,B7,B7      ;accum in B7
;branch occurs here
;cycles 110-124 (epilog)
                 ADDSP      .L1x   A7,B7,A0      ;lower/upper sum of products
                 ADDSP      .L2x   A7,B7,B0      ;
                 ADDSP      .L1x   A7,B7,A0      ;
                 ADDSP      .L2x   A7,B7,B0      ;
                 NOP                             ;wait for 1st B0
                 ADDSP      .L1x   A0,B0,A5      ;1st two sum of products
                 NOP                             ;wait for 2nd B0
                 ADDSP      .L2x   A0,B0,B5      ;last two sum of products
                 NOP               3             ;3 delay slots for ADDSP
                 ADDSP      .L1x   A5,B5,A4      ;final sum
                 NOP               3             ;3 delay slots for final sum
```

FIGURE 8.14. (*Continued*)

Cycle	Instruction	
110	ADDSP	
111	ADDSP	
112	ADDSP	
113	ADDSP	
114	NOP	
115	ADDSP	
116	NOP	
117	ADDSP	
118–120	NOP	3
121	ADDSP	
122–124	NOP	3

In cycles 113 through 116, A7 contains the lower 32-bit sum of products and B7 contains the upper 32-bit sum of products, or:

Cycle	A7 for Lower 32 bits (B7 for Upper 32 bits)
113	p0 + p4 + p8 + . . . + p96
114	p1 + p5 + p9 + . . . + p97
115	p2 + p6 + p10 + . . . + p98
116	p3 + p7 + p11 + . . . + p99

In cycle 114, A0 = A7 + B7 is available. A0 accumulates the lower and the upper sum of products, where

```
A7 = p0 + p4 + p8 + . . . + p96    (lower 32 bits)
B7 = p0 + p4 + p8 + . . . + p96    (upper 32 bits)
```

In cycle 115, B0 = A7 + B7 is available, where

```
A7 = p1 + p5 + p9 + . . . + p97    (lower 32 bits)
B7 = p1 + p5 + p9 + . . . + p97    (upper 32 bits)
```

Similarly, in cycles 116 and 117, A0 and B0 are obtained (available) as

```
A0 = sum of lower/upper 32 bits of (p2 + p6 + p10 + . . . + p98)
B0 = sum of lower/upper 32 bits of (p3 + p7 + p11 + . . . + p99)
```

In cycle 119, A5 = A0 + B0 (obtained from cycles 114 and 115). In cycle 121, B5 = A0 + B0 (obtained from cycles 116 and 117).

The final sum accumulates in A4 and is available after cycle 124.

8.6 EXECUTION CYCLES FOR DIFFERENT OPTIMIZATION SCHEMES

Table 8.4 shows a summary of the different optimization schemes for both fixed- and floating-point implementations, for a count of 200. The number of cycles can be obtained for different array sizes, since the number of cycles in the prolog and epilog stages remain the same.

Note that for a count of 1000, the fixed- and floating-point implementations with software pipeling take:

Fixed-point: 7 + (count/2) + 1 = 508 cycles
Floating-point: 9 + (count/2) + 15 = 524 cycles

TABLE 8.4 Number of Cycles with Different Optimization Schemes for Both Fixed- and Floating-Point Implementations (Count = 200)

Optimization Scheme	Number of Cycles	
	Fixed-Point	Floating-Point
No optimization	2 + (16 × 200) = 3202	2 + (18 × 200) = 3602
With parallel instructions	1 + (8 × 200) = 1601	1 + (10 × 200) = 2001
Two sums per iteration	1 + (8 × 100) = 801	1 + (10 × 100) + 7 = 1008
With software pipelining	7 + (100) + 1 = 108	9 + (100) + 15 = 124

REFERENCES

1. *TMS320C6000 Programmer's Guide,* SPRU198D, Texas Instruments, Dallas, TX, 2000.

2. *Guidelines for Software Development Efficiency on the TMS320C6000 VelociTI Architecture,* SPRA434, Texas Instruments, Dallas, TX, 1998.

3. *TMS320C6000 CPU and Instruction Set,* SPRU189F, Texas Instruments, Dallas, TX, 2000.

4. *TMS320C6000 Assembly Language Tools User's Guide,* SPRU186G, Texas Instruments, Dallas, TX, 2000.

5. *TMS320C6000 Optimizing Compiler User's Guide,* SPRU 187G, Texas Instruments, Dallas, TX, 2000.

9

DSP Applications and Student Projects

This chapter can be used as a source of experiments, projects, and applications, as well as Refs. 1 to 4. A wide range of projects have been implemented on the floating-point C30 and C31 processors [5–20] as well as on the fixed-point TMS320C25 [21–26]. They range in topics from communications and controls, to neural networks, and can be used as a source of ideas to implement other projects. The proceedings from the yearly conferences, published by Texas Instruments, contain a number of articles based on the TMS320 family of digital signal processors and can be a good source of project ideas. Texas Instruments' Web site contains a list of student projects covering a wide range of applications that have made it to the final rounds of the TI "DSP and Analog Design Contest Challenge" (which has a $100,000 first prize). Chapters 6 and 7 and Appendices D–F can also be useful.

I owe a special debt to all the students who have made this chapter possible. They include students from Roger Williams University and the University of Massachusetts–Dartmouth, who have contributed to my general background in DSP applications, in particular the Worcester Polytechnic Institute (WPI) students in my graduate course "Real-Time DSP," based on the C6x: Y. Bognadov, J. Boucher, G. Bowers, D. Ciota, P. DeBonte, B. Greenlaw, S. Kintigh, R. Lara-Montalvo, M. Mellor, F. Moyse, A. Pandey, I. Progri, V. C. Ramanna, P. Srikrishna, U. Ummethala, L. Wan. A brief discussion of their projects (and some miniprojects) are included in this chapter. Two projects on adaptive filtering and graphic equalizers were discussed in Chapters 6 and 7.

9.1 VOICE SCRAMBLER USING DMA AND USER SWITCHES (`scram16k_sw`)

The project **scram16k_sw** (on the accompanying disk) is an extension of Example 4.9, making use of the three dip switches, USER_SW1 through USER_SW3 (the fourth switch is not used), available on board the DSK. With voice as input, the output can be unscrambled voice (based on the user switch settings).

The user dip switches are used to determine whether or not to up-sample. The program can also be used as a loop or filter program, depending on the position of the switches. USER_SW1 corresponds to the LSB. A setting such as "down/down/up" represents $(001)_b$ and is the first one tested in the program. If true, the output is scrambled with up-sampling at 16 kHz. The following switch positions are used:

	USER_SW1	USER_SW2	USER_SW3	
a.	0	0	1	Output scrambled with $F_s = 16$ kHz
b.	1	0	1	Output unscrambled with $F_s = 16$ kHz
c.	1	1	1	Lowpass filtering with $F_s = 16$ kHz
d.	0	1	0	Output scrambled with $F_s = 8$ kHz
e.	1	1	0	Output unscrambled with $F_s = 8$ kHz
f.	0	0	0	Lowpass filtering with $F_s = 8$ kHz
g.	1	0	0	Loop program

scram8k_DMA

The alternative project **scram8k_DMA** (on the disk) implements the voice scrambling scheme using DMA, sampling at 8 kHz. It is adapted from the example codec_edma included with the DSK package. It illustrates the use of DMA with options within the program to inplement either a loop program, a filter, or the voice scrambling scheme (without up-sampling).

9.2 PHASE-LOCKED LOOP

The PLL project implements a software-based linear phase-locked loop (PLL). The basic PLL causes a particular system to track another PLL. It consists of a phase detector, a loop filter, and a voltage-controlled oscillator. The software PLL is more versatile. However, it is limited by the range in frequencies that can be covered, since the PLL function must be executed at least once every period of the input signal [27–29].

Initially, the PLL was tested using MATLAB, then ported to the C6x using C. The PLL locks to a sine wave, generated either internally within the program or from an external source. Output signals are viewed on a scope or on a PC using DSP/BIOS's real-time data transfer (RTDX).

Figure 9.1 shows a block diagram of the linear PLL, implemented in two versions:

1. Using an external input source, with the output of the digitally controlled oscillator (DCO) to an oscilloscope

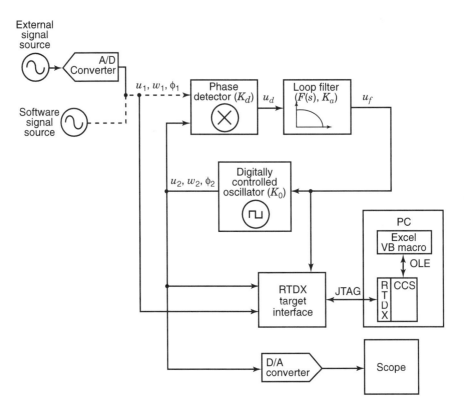

FIGURE 9.1. PLL block diagram.

2. Using RTDX with an input sine wave generated from a lookup table and various signals viewed using Excel

The phase detector, from Figure 9.1, multiplies the input sine wave by the square-wave output of the DCO. The sum and difference frequencies of the two inputs to the phase detector produces an output with a high- and a low-frequency component, respectively. The low-frequency component is used to control the loop, while the high-frequency component is filtered out. When the PLL is locked, the two inputs to the phase detector are at the same frequency but with a quadrature (90-degree) relationship.

The loop filter is a lowpass filter that passes the low-frequency output component of the phase detector while it attenuates the undesired high-frequency component. The loop filter is implemented as a single-pole IIR filter with a zero to improve the loop's dynamics and stability. The scaled output of the loop filter represents the instantaneous incremental phase step the DCO is to take. The DCO outputs a square wave as a Walsh function: +1 for phase between 0 and pi, and −1 for phase between −pi and 0; with incremental phase proportional to the number at its input.

9.2.1 RTDX for Real-Time Data Transfer

The RTDX feature was used to transfer data to the PC host using a sine wave from a lookup table as input. A single output channel was created to pass to CCS the input signal, the output of both the loop filter and the DCO, and time stamps. CCS buffers these data so that the data can be accessed by other applications on the PC host. CCS has an interface that allows PC applications to access buffered RTDX data. Visual Basic Excel was used (LABVIEW, or Visual C++ can also be used) to display the results on the PC monitor.

9.3 SB-ADPCM ENCODER/DECODER: IMPLEMENTATION OF G.722 AUDIO CODING

An audio signal is sampled at 16 kHz, transmitted at a rate of 64 kbits/s, and reconstructed at the receiving end [30,31].

Encoder

The subband adaptive differential pulse code modulated (SB-ADPCM) encoder consists of a transmit quadrature mirror filter that splits the input signal into a low-frequency band, 0 to 4 kHz, and a high-frequency band, 4 to 8 kHz. The low- and high-frequency signals are encoded separately by dynamically quantizing an adaptive predictor's output error. The low and the high encoder error signals are encoded with 6 and 2 bits, respectively. As long as the error signal is small, a negligible amount of overall quantization noise and good performance can be obtained. The low- and high-band bits are multiplexed and the result is 8 bits sampled at 8 kHz, for a bit rate of 64 kbits/s. Figure 9.2 shows a block diagram of a SB-ADPCM encoder.

Transmit Quadrature Mirror Filter

The transmit quadrature mirror filter (QMF) takes a 16-bit audio signal sampled at 16 kHz and separates it into a low band and a high band. The filter coefficients represent a 4-kHz lowpass filter. The sampled signal is separated into odd and even samples, with the effect of aliasing the signals from 4 to 8 kHz. This aliasing causes the high-frequency odd samples to be 180 degrees out of phase with the high-frequency even samples. The low-frequency even and odd samples are in-phase. When the odd and even samples are added, after being filtered, the low-frequency

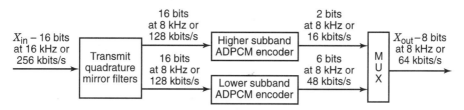

FIGURE 9.2. Block diagram of ADPCM encoder.

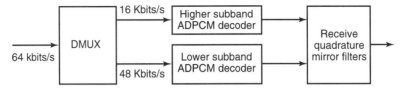

FIGURE 9.3. Block diagram of ADPCM decoder.

signals constructively add, while the high-frequency signals cancel each other, producing a low-band signal sampled at 8 kHz.

The low subband encoder converts the low frequencies from the QMF into an error signal that is quantized to 6 bits.

Decoder

The decoder decomposes a 64-kbits/s signal into two signals, to form the inputs to the lower and higher SB-ADPCM decoder, as shown in Figure 9.3. The receive quadrature mirror filter (QMF) consists of two digital filters to interpolate the lower- and higher-subband ADPCM decoders from 8 to 16 kHz and produce output at a rate of 16 kHz. In the higher SB-ADPCM decoder, adding the quantized difference signal to the signal estimate produces the reconstructed signal.

Components of the ADPCM decoder include an inverse adaptive quantizer, quantizer adaptation, adaptive prediction, predicted value computation, and reconstructed signal computation. With input from a CD player, the DSK reconstructed output signal sound quality was good. Buffered input and reconstructed output data also confirmed successful results from the decoder.

9.4 ADAPTIVE TEMPORAL ATTENUATOR

An adaptive temporal attenuator (ATA) suppresses undesired narrowband signals to achieve a maximum signal-to-interference ratio. Figure 9.4 shows a block diagram of the ATA. The input is passed through delay elements, and the outputs from selected delay elements are scaled by weights. The output is

$$y[k] = \mathbf{m}^T \cdot \mathbf{r}[k] = \sum_{i=0}^{N-1} (\mathbf{m}_i \cdot \mathbf{r}[k-i])$$

where $\mathbf{m}$ is a weight vector, $\mathbf{r}$ a vector of delayed samples selected from the input signal, and N the number of samples in $\mathbf{m}$ and $\mathbf{r}$. The adaptive algorithm computes the weights based on the correlation matrix and a direction vector:

$$\mathbf{C}[k, \delta = 0] \cdot \mathbf{m}[k] = \lambda \mathbf{D}$$

where $\mathbf{C}$ is a correlation matrix, $\mathbf{D}$ a direction vector, and λ a scale factor. The correlation matrix $\mathbf{C}$ is computed as an average of the signal correlation over several samples:

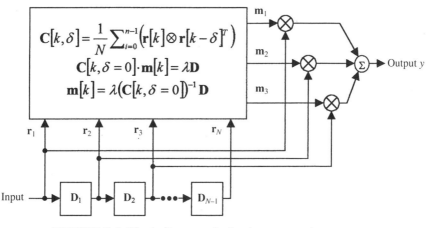

FIGURE 9.4. Block diagram of adaptive temporal attenuator.

$$C[k,\delta] = \frac{1}{N_{AV}}\sum_{i=0}^{n-1}\left(\mathbf{r}[k]\otimes\mathbf{r}[k-\delta]^T\right)$$

where N_{AV} is the number of samples included in the average. The direction vector **D** indicates the signal desired:

$$\mathbf{D} = [1 \quad \exp(j\omega_T\tau) \quad \cdots \quad \exp(j\omega_T(N-1)\tau)]^T$$

where ω_T is the angular frequency of the signal desired, τ the delay between samples that create the output, and N the order of the correlation matrix.

This procedure minimizes the undesired-to-desired ratio (UDR) [32]. UDR is defined as the ratio of the total signal power to the power of the signal desired, or

$$\text{UDR} = \frac{P_{\text{total}}}{P_d} = \frac{\mathbf{m}[k]^T \cdot \mathbf{C}[k,0] \cdot \mathbf{m}[k]}{P_d\left(\mathbf{m}[k]^T \cdot \mathbf{D}\right)^2} = \frac{1}{P_d\left(\mathbf{m}[k]^T \cdot \mathbf{D}\right)}$$

where P_d is the power of the signal desired.

MATLAB is used to simulate the ATA, then ported to the C6x for real-time implementation. Figure 9.5 shows the test setup using a fixed desired signal of 1416 Hz and an undesired signal of 1784 Hz (which can be varied). From MATLAB, an optimal value of τ is found to minimize UDR. This is confirmed in real time, since for that value of τ (varying τ with a GEL file), the undesired signal (initially displayed from an HP3561A analyzer) is greatly attenuated.

9.5 IMAGE PROCESSING

This project implements various schemes used in image processing:

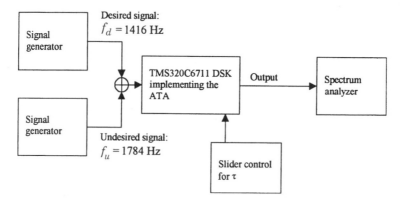

FIGURE 9.5. Test setup for adaptive temporal attenuator.

1. *Edge detection:* for enhancing edges in an image using Sobe's edge detection
2. *Median filtering:* nonlinear filter for removing noise spikes in an image
3. *Histogram equalization:* to make use of image spectrum
4. *Unsharp masking:* spatial filter to sharpen image, emphasizing high-frequency components of image
5. *Point detection:* for emphasizing single-point feature in image

A major issue was using/loading the images as .h files in lieu of using real-time images (due to the course one-semester time constraint). During the course of this project, the following evolved: a code example for additive noise with a Gaussian distribution, with adjustable variance and mean, and a code example on histogram transformation to map the distribution of one set of numbers to a different distribution (used in image processing).

9.6 FILTER DESIGN AND IMPLEMENTATION USING A MODIFIED PRONY'S METHOD

This project designs and implements a filter based on a modified Prony's method [33–36]. This method is based on the correlation property of the filter's representation and does not require computation of any derivatives or an initial guess of the coefficient vector. The filter's coefficients are calculated recursively to obtain the filter's impulse response.

9.7 FSK MODEM

This project implements a digital modulator/demodulator. It generates 8-ary FSK carrier tones. The following steps are performed in the program.

1. The sampled data are acquired as input.
2. The six most significant bits are separated into two 3-bit samples.
3. The most significant portion of the sample data selects an FSK tone.
4. The FSK tone is sent to a demodulator.
5. The FSK tone is windowed using the Hanning window function.
6. DFT (16-point) results are obtained for the windowed FSK tone.
7. DFT results are sent to the function that selects the frequency with the highest amplitude, corresponding to the upper 3 bits of the sampled data.
8. The process is repeated for the lower 3 bits of the sampled data.
9. The bits are combined and sent to the codec.
10. The gel program allows for an option to interpolate or up-sample the reconstructed data for a smoother output waveform.

9.8 µ-LAW FOR SPEECH COMPANDING

An analog input such as speech is converted into digital form and compressed into 8-bit data. µ-Law *encoding* is a nonuniform quantizing logarithmic compression scheme for audio signals. It is used in the United States to compress a signal into a logarithmic scale when coding for transmission. It is widely used in the telecommunications field because it improves the signal-to-noise ratio without increasing the amount of data.

The dynamic range increases while the number of bits for quantization remains the same. Typically, µ-law compressed speech is carried in 8-bit samples. It carries more information about smaller signals than about larger signals. It is based on the observation that many signals are statistically more likely to be near a low-signal level than a high-signal level. As a result, there are more quantization points nearer the low level.

A lookup table with 256 values is used to obtain the quantization levels from 0 to 7. The table consists of 16×16 set of numbers:

Two 0's
Two 1's
Four 2's
Eight 3's
Sixteen 4's
Thirty-two 5's
Sixty-four 6's
One hundred twenty-eight 7's

More of the higher-level signals are represented by 7 (from the lookup table). Three exponent bits are used to represent the levels from 0 to 7, four mantissa bits are used to represent the next four significant bits, and one bit is used for the sign bit.

The 16-bit input data is converted from linear to 8-bit μ-law (simulated for transmission), then converted back from μ-law to 16-bit linear (simulated as receiving), then output to the codec.

9.9 VOICE DETECTION AND REVERSE PLAYBACK

This project detects a voice signal from a microphone, then plays it back in the reverse direction. Two circular buffers are used; an input buffer to hold 80,000 samples (10 seconds of data) continuously being updated, and an output buffer to play back the input voice signal in the reverse direction. The signal level is monitored and its envelope is tracked to determine whether or not a voice signal is present.

When a voice signal appears and subsequently dies out, the signal-level monitor sends a command to start playback. The stored data are transferred from the input buffer to the output buffer for playback. Playback stops when reaching the end of the entire signal detected.

The signal-level monitoring scheme includes rectification and filtering (using a simple first-order IIR filter). An indicator specifies when the signal reaches an upper threshold. When the signal drops below a low threshold, the time difference between the start and end is calculated. If this time difference is less than a specified duration, the program continues into a no-signal state (if noise only). Otherwise, if it is more than a specified duration, a signal-detected mode is activated.

9.10 MISCELLANEOUS PROJECTS

The following projects were implemented using C/C3x and C2x/C5x code.

9.10.1 Acoustic Direction Tracker

The acoustic direction tracker has been implemented using C/C3x code and is discussed in Ref. 15. It uses two microphones to capture the signal. From the delay associated with the signal reaching one of the microphones before the other, a relative angle where the source is located can be determined. A signal radiated at a distance from its source can be considered to have a plane wavefront, as shown in Figure 9.6. This allows the use of equally spaced sensors (many microphones can be used as acoustical sensors) in a line to ascertain the angle at which the signal is radiating. Since one microphone is closer to the source than the other, the signal received by the more-distant microphone is delayed in time. This time shift corresponds to the angle where the source is located and the relative distance between the microphones and the source. The angle $c = \arcsin(a/b)$, where the distance a is the product of the speed of sound and the time delay (phase/frequency).

Figure 9.7 shows a block diagram of the acoustic signal tracker. Two 128-point

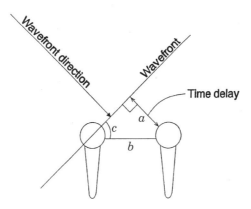

FIGURE 9.6. Signal reception with two microphones.

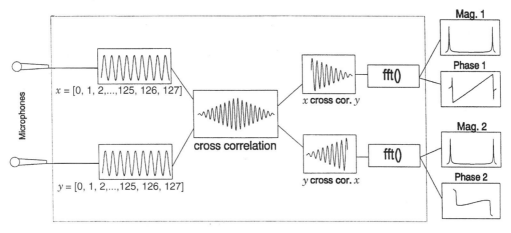

FIGURE 9.7. Block diagram of acoustic signal tracker.

arrays of data are obtained, cross-correlating the first signal with the second and then the second signal with the first. The resulting cross-correlation data are decomposed into two halves, each transformed using a 128-point FFT. The resulting phase is the phase difference of the two signals.

9.10.2 Multirate Filter

A filter can be realized with fewer coefficients using multirate processing, than with an equivalent single-rate approach. The multirate filter is discussed and implemented using C3x/C4x- and C2x/C5x-compatible code [37–44]. Possible applications include a graphic equalizer, a controlled noise source, and background noise synthesis. Multirate processing uses more than one sampling frequency to perform a desired processing operation. The two basic operations are decimation, which is a sampling-rate reduction, and interpolation, which is a sampling-rate increase [38–42]. Multirate decimators can reduce the computational requirements of the filter. A sampling-rate increase by a factor of K can be achieved with interpolation by padding (adding) $K - 1$ zeros between pairs of consecutive input samples x_i, x_{i+1}. Decimating or interpolating over several stages generally results in better efficiency.

A binary random signal is fed into a bank of filters that can be used to shape an output spectrum. Figure 9.8 shows a 10-band multirate filter discussed and implemented using C3x code [37] and C2x/C5x code [43,44]. The frequency range is divided into 10 octave bands, with each band being $\frac{1}{3}$-octave controllable.

9.10.3 Neural Network for Signal Recognition

The FFT of a signal becomes the input to a neural network, which is trained to recognize this input signal using a back-propagation learning rule [45,46] implemented in C. A three-layer neural network using seven nodes (Figure 9.9) was used to illustrate the algorithm. Many different rules are available for training a neural network, and back-propagation has been used for a wide range of applications. Given a set of inputs, the network is trained to give a desired response. If the network gives the wrong answer, the network is corrected by adjusting its parameters (weights) so that the error is reduced. During this correction process, one starts with the output nodes and propagation is backward to the input nodes.

9.10.4 PID Controller

Both nonadaptive and adaptive controllers using proportional, integral, and derivative (PID) control algorithm have been implemented in Refs. 6, 47, and 48.

9.10.5 Four-Channel Multiplexer for Fast Data Acquisition

A four-channel multiplexer module was designed and built for this project, implemented in C [6]. It includes an 8-bit flash ADC, a FIFO, a MUX, and a crystal oscillator (2 or 20 MHz). An input is acquired through one of the four channels. The FFT of the input signal is displayed in real time on the PC monitor.

9.10.6 Video Line Rate Analysis

This project is discussed in Refs. 6 and 49 and implemented using C/C3x code. It analyzes a video signal at the horizontal (line) rate. Interactive algorithms commonly used in image processing for filtering, averaging, and edge enhancement using C code are utilized for this analysis. The source of the video signal is a charge-coupled device (CCD) camera as input to a module designed and built for this

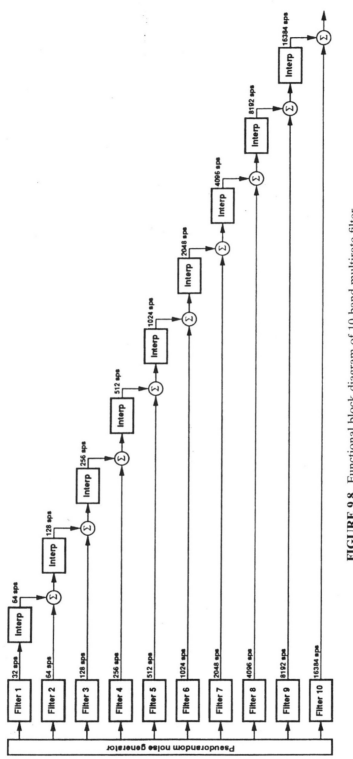

FIGURE 9.8. Functional block diagram of 10-band multirate filter.

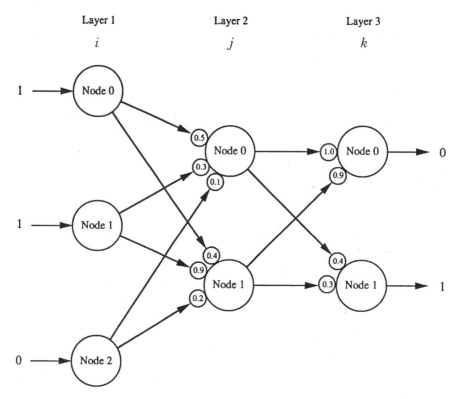

FIGURE 9.9. Three-layer neural network with seven nodes.

project. This module include flip-flops, logic gates, and a clock. Displays on the PC monotor illustrate various effects on one horizontal video line signal from either a 500-kHz or a 3-MHz IIR lowpass filter and from an edge enhancement algorithm.

REFERENCES

1. J. H. McClellan, R. W. Schafer, and M. A. Yoder, *DSP First: A Multimedia Approach*, Prentice Hall, Upper Saddle River, NJ, 1998.

2. N. Kehtarnavaz and M. Keramat, *DSP System Design Using the TMS320C6000*, Prentice Hall, Upper Saddle River, NJ, 2001.

3. N. Dahnoun, *DSP Implementation Using the TMS320C6x Processors*, Prentice Hall, Upper Saddle River, NJ, 2000.

4. M. Morrow, T. Welch, C. Cameron, and G. York, Teaching real-time beamforming with the C6211 DSK and MATLAB, *Proceedings of the Texas Instruments DSPS Fest Annual Conference*, 2000.

5. R. Chassaing, *Digital Signal Processing Laboratory Experiments Using C and the TMS320C31 DSK*, Wiley, New York, 1999.

6. R. Chassaing, *Digital Signal Processing with C and the TMS320C30*, Wiley, New York, 1992.

7. C. Marven and G. Ewers, *A Simple Approach to Digital Signal Processing*, Wiley, New York, 1996.

8. J. Chen and H. V. Sorensen, *A Digital Signal Processing Laboratory Using the TMS320C30*, Prentice Hall, Upper Saddle River, NJ, 1997.

9. S. A. Tretter, *Communication System Design Using DSP Algorithms*, Plenum Press, New York, 1995.

10. R. Chassaing et al., Student projects on digital signal processing with the TMS320C30, *Proceedings of the 1995 ASEE Annual Conference*, June 1995.

11. J. Tang, Real-time noise reduction using the TMS320C31 digital signal processing starter kit, *Proceedings of the 2000 ASEE Annual Conference*, 2000.

12. C. Wright, T. Welch III, M. Morrow, and W. J. Gomes III, Teaching real-world DSP using MATLAB and the TMS320C31 DSK, *Proceedings of the 1999 ASEE Annual Conference*, 1999.

13. J. W. Goode and S. A. McClellan, Real-time demonstrations of quantization and prediction using the C31 DSK, *Proceedings of the 1998 ASEE Annual Conference*, 1998.

14. R. Chassaing and B. Bitler (contributors), Signal processing chips and applications, *The Electrical Engineering Handbook*, CRC Press, Boca Raton, FL, 1997.

15. R. Chassaing et al., Digital signal processing with C and the TMS320C30: Senior projects, *Proceedings of the 3rd Annual TMS320 Educators Conference*, Texas Instruments, Dallas, TX, 1993.

16. R. Chassaing et al., Student projects on applications in digital signal processing with C and the TMS320C30, *Proceedings of the 2nd Annual TMS320 Educators Conference*, Texas Instruments, Dallas, TX, 1992.

17. R. Chassaing, TMS320 in a digital signal processing lab, *Proceedings of the TMS320 Educators Conference*, Texas Instruments, Dallas, TX, 1991.

18. P. Papamichalis, ed., *Digital Signal Processing Applications with the TMS320 Family: Theory, Algorithms, and Implementations*, Vols. 2 and 3, Texas Instruments, Dallas, TX, 1989 and 1990.

19. *Digital Signal Processing Applications with the TMS320C30 Evaluation Module: Selected Application Notes*, Texas Instruments, Dallas, TX, 1991.

20. R. Chassaing and D. W. Horning, *Digital Signal Processing with the TMS320C25*, Wiley, New York, 1990.

21. I. Ahmed, ed., *Digital Control Applications with the TMS320 Family*, Texas Instruments, Dallas, TX, 1991.

22. A. Bateman and W. Yates, *Digital Signal Processing Design*, Computer Science Press, New York, 1991.

23. Y. Dote, *Servo Motor and Motion Control Using Digital Signal Processors*, Prentice Hall, Upper Saddle River, NJ, 1990.

24. R. Chassaing, A senior project course in digital signal processing with the TMS320, *IEEE Transactions on Education*, Vol. 32, 1989, pp. 139–145.

25. R. Chassaing, Applications in digital signal processing with the TMS320 digital signal processor in an undergraduate laboratory, *Proceedings of the 1987 ASEE Annual Conference*, June 1987.

26. K. S. Lin, ed., *Digital Signal Processing Applications with the TMS320 Family: Theory, Algorithms, and Implementations*, Prentice Hall, Upper Saddle River, NJ, Vol. 1, 1988.

27. Roland E. Best, *Phase-Locked Loops Design, Simulation, and Applications*, 4th ed., McGraw-Hill, New York, 1999.

28. W. Li and J. Meiners, *Introduction to Phase-Locked Loop System Modeling*, SLTT015, Texas Instruments, Dallas, TX, May 2000.

29. J. P. Hein and J. W. Scott, Z-domain model for discrete-time PLL's, *IEEE Transactions on Circuits and Systems*, Vol. CS-35, Nov. 1988, pp. 1393–1400.

30. *ITU-T Recommendation G.722 Audio Coding with 64 kbits/s.*

31. P. M. Embree, *C Algorithms for Real-Time DSP*, Prentice Hall, Upper Saddle River, NJ, 1995.

32. I. Progri and W. R. Michalson, Adaptive spatial and temporal selective attenuator in the presence of mutual coupling and channel errors, *ION GPS-2000*.

33. F. Brophy and A. C. Salazar, Recursive digital filter synthesis in the time domain, *IEEE Transactions on Acoustics, Speech, and Signal Processing*, Vol. ASSP-22, 1974.

34. W. H. Press, S. A. Teukolsky, W. T. Vetterling, and B. P. Flannery, *Numerical Recipes in C: The Art of Scientific Computing*, Cambridge University Press, New York, 1992.

35. J. Borish and J. B. Angell, An efficient algorithm for measuring the impulse response using pseudorandom noise, *Journal of the Audio Engineering Society*, Vol. 31, 1983.

36. T. W. Parks and C. S. Burrus, *Digital Filter Design*, Wiley, New York, 1987.

37. R. Chassaing, P. Martin, and R. Thayer, Multirate filtering using the TMS320C30 floating-point digital signal processor, *Proceedings of the 1991 ASEE Annual Conference*, June 1991.

38. R. E. Crochiere and L. R. Rabiner, *Multirate Digital Signal Processing*, Prentice Hall, Upper Saddle River, NJ, 1983.

39. R. W. Schafer and L. R. Rabiner, A digital signal processing approach to interpolation, *Proceedings of the IEEE*, Vol. 61, 1973, pp. 692–702.

40. R. E. Crochiere and L. R. Rabiner, Optimum FIR digital filter implementations for decimation, interpolation and narrow-band filtering, *IEEE Transactions on Acoustics, Speech, and Signal Processing*, Vol. ASSP-23, 1975, pp. 444–456.

41. R. E. Crochiere and L. R. Rabiner, Further considerations in the design of decimators and interpolators, *IEEE Transactions on Acoustics, Speech, and Signal Processing*, Vol. ASSP-24, 1976, pp. 296–311.

42. M. G. Bellanger, J. L. Daguet, and G. P. Lepagnol, Interpolation, extrapolation, and reduction of computation speed in digital filters, *IEEE Transactions on Acoustics, Speech, and Signal Processing*, Vol. ASSP-22, 1974, pp. 231–235.

43. R. Chassaing, W. A. Peterson, and D. W. Horning, A TMS320C25-based multirate filter, *IEEE Micro*, Oct. 1990, pp. 54–62.

44. R. Chassaing, Digital broadband noise synthesis by multirate filtering using the TMS320C25, *Proceedings of the 1988 ASEE Annual Conference*, Vol. 1, June 1988.

45. B. Widrow and R. Winter, Neural nets for adaptive filtering and adaptive pattern recognition, *Computer*, Mar. 1988, pp. 25–39.

46. D. E. Rumelhart, J. L. McClelland, and the PDP Research Group, *Parallel Distributed Processing: Explorations in the Microstructure of Cognition*, Vol. 1, MIT Press, Cambridge, MA, 1986.

47. J. Tang, R. Chassaing, and W. J. Gomes III, Real-time adaptive PID controller using the TMS320C31 DSK *Proceedings of the 2000 Texas Instruments DSPS Fest Conference*, 2000.

48. J. Tang and R. Chassaing, PID controller using the TMS320C31 DSK for real-time motor control, *Proceedings of the 1999 Texas Instruments DSPS Fest Conference*, 1999.

49. B. Bitler and R. Chassaing, Video line rate processing with the TMS320C30, *Proceedings of the 1992 International Conference on Signal Processing Applications and Technology (ICSPAT)*, 1992.

50. *MATLAB, The Language of Technical Computing, Version 6.3*, MathWorks, Natick, MA, 1999.

A

TMS320C6x Instruction Set

A.1 INSTRUCTIONS FOR FIXED- AND FLOATING-POINT OPERATIONS

Table A.1 shows a listing of the instructions available for the C6x processors. The instructions are grouped under the functional units used by these instructions. These instructions can be used with both fixed- and floating-point C6x processors.

A.2 INSTRUCTIONS FOR FLOATING-POINT OPERATIONS

Table A.2 shows a listing of additional instructions available with the floating-point processor C67x. These instructions handle floating-point type of operations and are grouped under the functional units used by these instructions (see also Table A.1).

REFERENCES

1. *C6000 CPU and Instruction Set*, SPRU189F, Texas Instruments, Dallas, TX, 2000.
2. *TMS320 TMS320C6000 Programmer's Guide*, SPRU198D, Texas Instruments, Dallas, TX, 2000.

TABLE A.1   Instructions for Fixed- and Floating-Point Operations

.L Unit	.M Unit	.S Unit	.D Unit
ABS	MPY	ADD	ADD
ADD	MPYH	ADDK	ADDAB
ADDU	MPYHL	ADD2	ADDAH
AND	MPYHLU	AND	ADDAW
CMPEQ	MPYHSLU	B disp	LDB
CMPGT	MPYHSU	B IRP[a]	LDBU
CMPGTU	MPYHU	B NRP[a]	LDH
CMPLT	MPYHULS	B reg	LDHU
CMPLTU	MPYHUS	CLR	LDW
LMBD	MPYLH	EXT	LDB (15-bit offset)[b]
MV	MPYLHU	EXTU	LDBU (15-bit offset)[b]
NEG	MPYLSHU	MV	LDH (15-bit offset)[b]
NORM	MPYLUHS	MVC[a]	LDHU (15-bit offset)[b]
NOT	MPYSU	MVK	LDW (15-bit offset)[b]
OR	MPYU	MVKH	MV
SADD	MPYUS	MVKLH	STB
SAT	SMPY	NEG	STH
SSUB	SMPYH	NOT	STW
SUB	SMPYHL	OR	STB (15-bit offset)[b]
SUBU	SMPYLH	SET	STH (15-bit offset)[b]
SUBC		SHL	STW (15-bit offset)[b]
XOR		SHR	SUB
ZERO		SHRU	SUBAB
		SSHL	SUBAH
		SUB	SUBAW
		SUBU	ZERO
		SUB2	
		XOR	
		ZERO	

[a]S2 only.   [b]D2 only.

Source: Courtesy of Texas Instruments [1,2].

TABLE A.2   Instructions for Floating-Point Operations

.L Unit	.M Unit	.S Unit	.D Unit
ADDDP	MPYDP	ABSDP	ADDAD
ADDSP	MPYI	ABSSP	LDDW
DPINT	MPYID	CMPEQDP	
DPSP	MPYSP	CMPEQSP	
DPTRUNC		CMPGTDP	
INTDP		CMPGTSP	
INTDPU		CMPLTDP	
INTSP		CMPLTSP	
INTSPU		RCPDP	
SPINT		RCPSP	
SPTRUNC		RSQRDP	
SUBDP		RSQRSP	
SUBSP		SPDP	

Source: Courtesy of Texas Instruments [1,2].

B

Registers for Circular Addressing and Interrupts

A number of special-purpose registers available on the C6x processor are shown in Figures B.1 to B.8.

1. Figure B.1 shows the address mode register (AMR) that is used for the circular mode of addressing. It is used to select one of eight register pointers (A4 through A7, B4 through B7), and two blocks of memories (BK0, BK1) that can be used as circular buffers.
2. Figure B.2 shows the control status register (CSR) with bit 0 for the global interrupt enable (GIE) bit.
3. Figure B.3 shows the interrupt enable register (IER).
4. Figure B.4 shows the interrupt flag register (IFR).
5. Figure B.5 shows the interrupt set register (ISR).
6. Figure B.6 shows the interrupt clear register (ICR).
7. Figure B.7 shows the interrupt service table pointer (ISTP).
8. Figure B.8 shows the serial port control register (SPCR).

In Section 3.7.2 we discuss the AMR register and in Section 3.14 the interrupt registers.

REFERENCE

1. *C6000 CPU and Instruction Set*, SPRU189F, Texas Instruments, Dallas, TX, 2000.

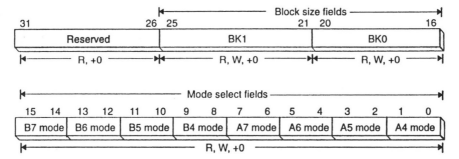

FIGURE B.1. Address mode register (AMR). (Courtesy of Texas Instruments.)

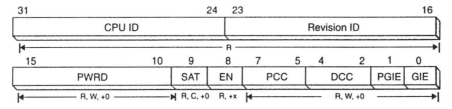

FIGURE B.2. Control status register (CSR). (Courtesy of Texas Instruments.)

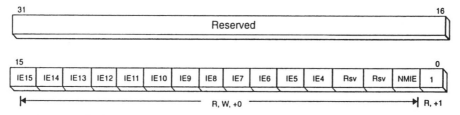

FIGURE B.3. Interrupt enable register (IER). (Courtesy of Texas Instruments.)

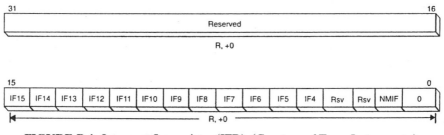

FIGURE B.4. Interrupt flag register (IFR). (Courtesy of Texas Instruments.)

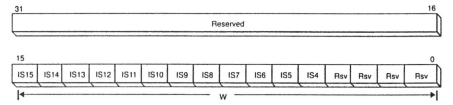

FIGURE B.5. Interrupt set register (ISR). (Courtesy of Texas Instruments.)

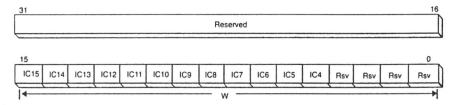

FIGURE B.6. Interrupt clear register (ICR). (Courtesy of Texas Instruments.)

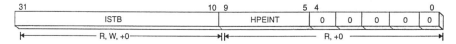

FIGURE B.7. Interrupt service table pointer (ISTP). (Courtesy of Texas Instruments.)

31		24	23	22	21	20	19	18	17	16
reserved†			F̄R̄S̄T̄	Ḡ R̄ S̄ T̄	XINTM		XSYNCERR‡	X̄ĒM̄P̄T̄Ȳ	XRDY	X̄R̄S̄T̄
R, +0			RW, +0	RW, +0	RW, +0		RW, +0	R, +0	R, +0	RW, +0

15	14	13	12	10	9	8	7	6	5	4	3	2	1	0
DLB	RJUST	CLKSTP	reserved				reserved	reserved	RINTM		RSYNCERR	RFULL	RRDY	R̄R̄S̄T̄
RW,+0	RW, +0	RW,+0	R, +0				R, +0	R, +0	RW, +0		RW, +0	R, +0	R, +0	RW, +0§

FIGURE B.8. Serial port control register (SPCR). (Courtesy of Texas Instruments.)

C

Fixed-Point Considerations

The C6711 is a floating-point processor capable of performing both integer and floating-point operations. Both the C6711 and the AD535 codec support 2's-complement arithmetic. It is thus appropriate here to review some fixed-point concepts [1].

In a fixed-point processor, numbers are represented in integer format. In a floating-point processor, both fixed- and floating-point arithmetic can be handled. With the floating-point processor C6711, a much greater range of numbers can be represented than with a fixed-point processor.

The dynamic range of an N-bit number based on 2's-complement representation is between $-(2^{N-1})$ and $(2^{N-1} - 1)$, or between $-32,768$ and $32,767$ for a 16-bit system. By normalizing the dynamic range between -1 and 1, the range will have 2^N sections, where $2^{-(N-1)}$ is the size of each section starting at -1 up to $1 - 2^{-(N-1)}$. For a 4-bit system, there would be 16 sections, each of size $^1/_8$, from -1 to $^7/_8$.

C.1 BINARY AND TWO'S-COMPLEMENT REPRESENTATION

To make illustrations more manageable, a 4-bit system is used rather than a 32-bit word length. A 4-bit word can represent the unsigned numbers 0 through 15, as shown in Table C.1.

The 4-bit unsigned numbers represent a modulo (mod) 16 system. If 1 is added to the largest number (15), the operation wraps around to give 0 as the answer. Finite bit systems have the same modulo properties as do number wheels on combination locks. Therefore, a number wheel graphically demonstrates the addition properties of a finite bit system. Figure C.1 shows a number wheel with the numbers 0 through 15 wrapped around the outside. For any two numbers x and y in the range, the operation amounts to the following procedure:

TABLE C.1   Unsigned Binary Number

Binary	Decimal
0000	0
0001	1
0010	2
0011	3
.	.
.	.
.	.
1110	14
1111	15

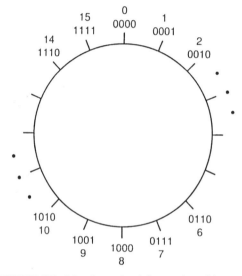

FIGURE C.1. Number wheel for unsigned integers.

1. Find the first number x on the wheel.
2. Step off y units in the clockwise direction, which brings you to the answer.

For example, consider the addition of the two numbers $(5 + 7)$ mod 16, which yields 12. From the number wheel, locate 5, then step 7 units in the clockwise direction to arrive at the answer, 12. As another example, $(12 + 10)$ mod16 = 6. Starting with 12 on the number wheel, step 10 units clockwise, past zero, to 6.

Negative numbers require a different interpretation of the numbers on the wheel. If we draw a line through 8 cutting the number wheel in half, the right half will represent the positive numbers and the left half the negative numbers, as shown in Figure C.2. This representation is the 2's-complement system. The negative numbers are the 2's complement of the positive numbers, and vice versa.

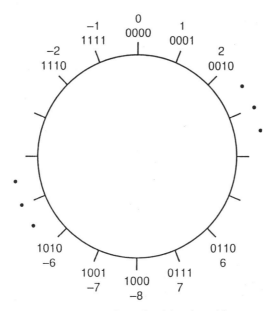

FIGURE C.2. Number wheel for signed integers.

A 2's-complement binary integer,

$$B = b_{n-1} \cdots b_1 b_0$$

is equivalent to the decimal integer

$$I(B) = -b_{n-1} \times 2^{n-1} + \cdots + b_1 \times 2^1 + b_0 \times 2^0$$

where the b's are binary digits. The sign bit has a negative weight; all the others have positive weights. For example, consider the number –2,

$$1110 = -1 \times 2^3 + 1 \times 2^2 + 1 \times 2^1 + 0 \times 2^0 = -8 + 4 + 2 + 0 = -2$$

To apply the graphical technique to the operation 6 + (–2) mod16 = 4, locate 6 on the wheel, then step off (1110) units clockwise to arrive at the answer 4.

The binary addition of these same numbers,

$$
\begin{array}{r}
0110 \\
1110 \\
\hline
10100 \\
C
\end{array}
$$

shows a carry in the most significant bit, which in the case of a finite register arithmetic, will be ignored. This carry corresponds to the wraparound through zero on

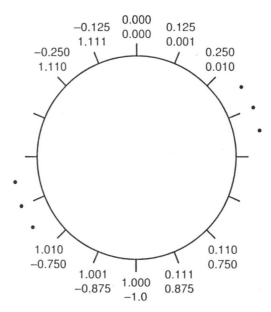

FIGURE C.3. Number wheel for fixed-point representation.

the number wheel. The addition of these two numbers results in correct answers, by ignoring the carry in the most significant bit position, provided that the answer is in the range of representable numbers -2^{n-1} to $(2^{n-1} - 1)$ in the case of an n-bit number, or between -8 and 7 for the 4-bit number wheel example. When -7 is added to -8 in the 4-bit system, we get an answer of $+1$ instead of the correct value of -15, which is out of range. When two numbers of like sign are added to produce an answer with opposite sign, overflow has occurred. Subtraction with 2's-complement numbers is equivalent to adding the 2's complement of the number being subtracted to the other number.

C.2 FRACTIONAL FIXED-POINT REPRESENTATION

Rather than using the integer values just discussed, a fractional fixed-point number that has values between $+0.99\ldots$ and -1 can be used. To obtain the fractional n-bit number, the radix point must be moved $n - 1$ places to the left. This leaves one sign bit plus $n - 1$ fractional bits. The expression

$$F(B) = -b_0 \times 2^0 + b_1 \times 2^{-1} + b_2 \times 2^{-2} + \cdots + b_{n-1} \times 2^{-(n-1)}$$

converts a binary fraction to a decimal fraction. Again, the sign bit has a weight of negative 1 and the weights of the other bits are positive powers of $\frac{1}{2}$. The number wheel representation for the fractional 2's-complement 4-bit numbers is shown in Figure C.3. The fractional numbers are obtained from the 2's-complement integer numbers of Figure C.2 by scaling them by 2^3. Because the number of bits in a 4-bit system is small, the range is from −1 to 0.875. For a 16-bit word, the signed integers range from −32,768 to +32,767. To get the fractional range, scale those two signed integers by 2^{-15} or 32,768, which results in a range from −1 to 0.999969 (usually taken as 1).

C.3 MULTIPLICATION

If one multiplies two n-bit numbers, the common notion is that a $2n$-bit operand will result. Although this is true for unsigned numbers, it is not so for signed numbers. As shown before, sign numbers need one sign bit with a weight of -2^{n-1}, followed by positive weights that are powers of 2. To find the number of bits needed for the result, multiply the two largest numbers together:

$$P = (-2^{n-1})(-2^{n-1}) = 2^{2n-2}$$

This number is a positive number representable in $(2n - 1)$ bits. The most significant bit of this result occupies the $(2n - 2)$ bit position counting from 0. Since this number is positive, its sign bit, which would show up as a negative number (a power of 2), does not appear. This is an exceptional case, which is treated as an overflow in fractional representation. Since the fractional representation requires that both operand and resultant occupy the same range, $-1 \geqslant$ range $< +1$, the operation $(-1) \times (-1)$ produces an unrepresentable number, +1.

Consider the next larger combination:

$$P = (-2^{n-1})(-2^{n-1} + 1) = 2^{2n-2} - 2^{n-1}$$

Since the second number subtracts from the first, the product will occupy up to the $(2n - 3)$ bit position, counting from 0. Thus, it is representable in $(2n - 2)$ bits. With the exceptional case ruled out, this makes the bit position $(2n - 2)$ available for the sign bit of the resultant. Therefore, $(2n - 1)$ bits are needed to support an $(n \times n)$-bit signed multiplication.

To clarify the preceding equation, consider the 4-bit case, or

$$P = (-2^3)(-2^3 + 1) = 2^6 - 2^3$$

The number 2^6 occupies bit position 6. Since the second number is negative, the summation of the two is a number that will occupy only bit positions less than bit position 6, or

$$2^6 - 2^3 = 64 - 8 = 56 = 00111000$$

Thus bit position 6 is available for the sign bit. The 8-bit equivalent would have two sign bits (bits 6 and 7). The C6x supports signed and unsigned multiplies and therefore provides $2n$ bits for the product.

Consider the multiplication of two fractional 4-bit numbers, with each number consisting of 3 fractional bits and 1 sign bit. Let the product be represented by an 8-bit number. The first number is −0.5 and the second number is 0.75; the multiplication is as follows:

$$-0.50 = 1.100$$
$$\times 0.75 = 0.110$$

$$\underline{1111}1000$$
$$\underline{111}000$$

$$111.101000$$
$$\text{C}$$
$$= -2^1 + 2^0 + 2^{-1} + 2^{-3} = -0.375$$

The underlined bits of the multiplicand indicate sign extension. When a negative multiplicand is added to the partial product, it must be sign-extended to the left up to the limit of the product, in order to give the proper larger bit version of the same number. To demonstrate that sign extension gives the correct expanded bit number, scan around the number wheel in Figure C.2 in the counterclockwise direction from 0. Write the codes for 5-bit, 6-bit, 7-bit, . . . negative numbers. Notice that they would be derived correctly by sign-extending the existing 4-bit codes; therefore, sign extension gives the correct expanded bit number. The carry-out will be ignored; however, the numbers 111.101000 (9-bit word), 11.101000 (8-bit word), and 1.101000 (7-bit word) all represent the same number: −0.375. Thus, the product of the preceding example could be represented by $(2n − 1)$ bits, or 7 bits for a 4-bit system.

When two 16-bit numbers are multiplied to produce a 32-bit result, only 31 bits are needed for the multiply operation. As a result, bit 30 is sign-extended to bit 31. The extended bits are frequently called *sign bits*.

Consider the following example: to multiply $(0101)_2$ by $(1110)_2$, which is equivalent to multiplying 5 by −2 in decimal, which would result in −10. This result is outside the dynamic range {−8,7} of a 4-bit system. Using a Q-3 format, this corresponds to multiplying 0.625 by −0.25, yielding a result of −0.15625, which is within the fractional range.

When two Q-15 format numbers (each with a sign bit) are multiplied, the result is a Q-30 format number with one extra sign bit. The most significant bit is the extra sign bit. One can shift right by 15 to retain the most significant bits and only one of the two sign bits. By shifting right by 15 (dividing by 2^{15}) to be able to store the result into a 16-bit system, this discards the 15 least significant bits, thereby losing some precision. One is able to retain high precision by keeping the most significant 15 bits. With a 32-bit system, a left shift by one bit would suffice to get rid of the extra sign bit.

Note that when two Q-15 numbers, represented with a range of −1 to 1, are multiplied, the resulting number remains within the same range. However, the addition of two Q-15 numbers can produce a number outside this range, causing overflow. Scaling would then be required to correct this overflow.

Since the AD535 is a 16-bit system, a 32-bit result must eventually be truncated or rounded to 16 bits. The most significant bits, along with the sign bit and its duplicate, are in the high end of the accumulating 32-bit register of the C6x. The result in the high end of the accumulating register is left-shifted to eliminate the extra sign bit and to give an additional bit of resolution when moved to a 16-bit location.

REFERENCE

1. R. Chassaing and D. W. Horning, *Digital Signal Processing with the TMS320C25*, Wiley, New York, 1990.

D

MATLAB Support Tools

Several support tools using MATLAB [1,2] are described in this appendix:

1. Filter designer SPTOOL for FIR and IIR filter design using a graphical user interface (GUI); RTSPTOOL as an extension to SPTOOL
2. FIR and IIR filter design using functions available with the Student Version of MATLAB
3. Bilinear transformation
4. FFT and IFFT

D.1 MATLAB GUI FILTER DESIGNER SPTOOL FOR FIR FILTER DESIGN

MATLAB provides a graphical user interface (GUI) filter designer SPTOOL for the design of FIR (and IIR) filters.

Example D.1: MATLAB GUI Filter Designer SPTOOL for FIR Filter Design

1. From MATLAB, type the following:

   ```
   >>sptool
   ```

   to access MATLAB's GUI filter designer SPTOOL for the design of both FIR and IIR filters.
2. From the startup window startup.spt, select a new design and use the characteristics shown in Figure D.1 to design an FIR bandstop filter centered at

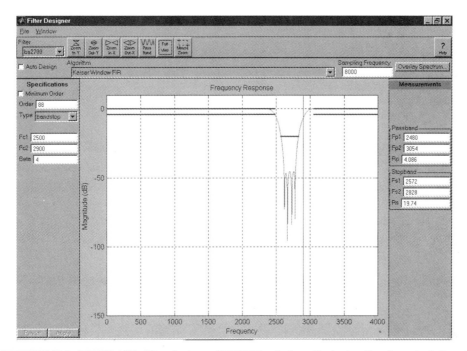

FIGURE D.1. MATLAB's filter designer SPTOOL window displaying the characteristics of an FIR bandstop filter centered at 2700 Hz.

2700 Hz. The filter contains $N = 89$ coefficients (MATLAB shows order as $N - 1$) and uses the Kaiser window function. The real-time implementation of this filter is tested in Example 4.1.

3. When finished, access the `startup` window again. Select → Edit → Name. Change name (enter new variable name) to *bs2700*.

4. Select File → Export → Export to Workspace the *bs2700* design.

5. Access MATLAB's workspace and type the following two commands:

```
>>bs2700.tf.num;
>>round(bs2700.tf.num*2^15)
```

to find the numerator coefficients of the transfer function, and scale them by 2^{15}. The scaled coefficients of the FIR bandstop filter should be listed within the workspace as

```
-14  23  -9 . . . 23  -14
```

These coefficients are contained in the file *bs2700.cof*, shown in Figure D.2 and used in Example 4.1.

```
//BS2700.cof  FIR bandstop coefficients designed with MATLAB

#define N 89                              //number of coefficients

short h[N]={-14,23,-9,-6,0,8,16,-58,50,44,-147,119,67,-245,200,72,
-312,257,53,-299,239,20,-165,88,0,105,-236,33,490,-740,158,932,-1380,
392,1348,-2070,724,1650,-2690,1104,1776,-3122,1458,1704,29491,1704,
1458,-3122,1776,1104,-2690,1650,724,-2070,1348,392,-1380,932,158,-740,
490,33,-236,105,0,88,-165,20,239,-299,53,257,-312,72,200,-245,67,119,
-147,44,50,-58,16,8,0,-6,-9,23,-14};
```

FIGURE D.2. Coefficient file for an FIR bandstop filter centered at 2700 Hz designed using MATLAB's filter designer SPTOOL (bs2700.cof).

Real-Time SPTOOL (RTSPTOOL)

Real-time SPTOOL (RTSPTOOL) provides a direct interface for the DSK [3–5] for filter design and implementation (within the MATLAB's environment) on the DSK in real time. RTSPTOOL's window is similar to SPTOOL's filter designer window, with additional toolbars to run the filter in real time on the DSK. Upon pressing an appropriate toolbar, the filter is designed and the coefficients are scaled and saved into an appropriate file that is included in a generic FIR program. MATLAB's file *filtdes.m* was modified to provide that interface to the DSK. A (MATLAB .m) function accesses CCS code generation tools to compile/assemble, link, and load/run the resulting executable file on the DSK (load/run using *dsk6xldr filename.out*).

D.2 MATLAB GUI FILTER DESIGNER SPTOOL FOR IIR FILTER DESIGN

Section D.1 illustrates the design of FIR filters using MATLAB's GUI filter designer SPTOOL. Some of the same procedures are used for the design of IIR filters as well.

Example D.2: MATLAB GUI Filter Designer SPTOOL for IIR Filter Design

Figure D.3 shows MATLAB's filter designer SPTOOL displaying the characteristics of a tenth-order IIR bandstop filter centered at 1750 Hz. MATLAB shows the order as 5, which represents the number of second-order sections. Save it as *bs1750* (see Example D.1). Export the coefficients to the workspace as with the previous FIR design. From MATLAB's workspace, type the following commands:

```
>>[z,p,k] = tf2zp(bs1750.tf.num, bs1750.tf.den);
>>sec_ord_sec = zp2sos(z,p,k);
>>sec_ord_sec = round(sec_ord_sec*2^15)
```

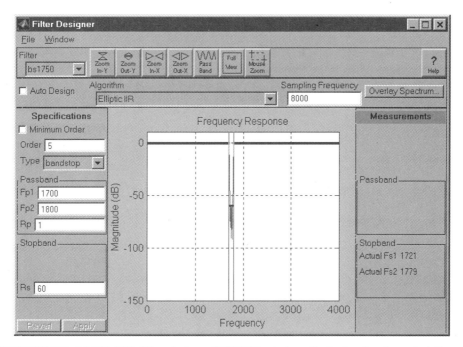

FIGURE D.3. MATLAB's filter designer SPTOOL window displaying the characteristics of an IIR bandstop filter centered at 1750 Hz.

The first command finds the roots of the numerator and the denominator (zeros and poles) and converts the results (scaled) into a format for implementation as second-order sections. The resulting numerator and denominator coefficients should be listed as

```
27940    -10910    27940    32768    -11417    25710
.
.
.
32768    -14239    32768    32768    -15258    32584
```

These 30 coefficients represent the numerator coefficients a_0, a_1, and a_2 and the denominator coefficients b_0, b_1, and b_2. They represent six coefficients per stage, with b_0 normalized to 1 and scaled by $2^{15} = 32{,}768$. These coefficients are contained in the file *bs1750.cof*, listed in Figure D.4 and used in Example 5.1. Figure D.4 shows 25 coefficients (in lieu of 30). Since the coefficient b_0 is always normalized to 1, it is not used in the program. As with the FIR design, this IIR bandstop filter can be implemented in real time with a push of a button within RTSPTOOL [3,4].

```
//bs1750.cof IIR bandstop coefficient file, centered at 1,750 Hz

#define stages 5                   //number of 2nd-order stages

int a[stages][3]=        {        //numerator coefficients
{27940, -10910, 27940},           //a10, a11, a12 for 1st stage
{32768, -11841, 32768},           //a20, a21, a22 for 2nd stage
{32768, -13744, 32768},           //a30, a31, a32 for 3rd stage
{32768, -11338, 32768},           //a40, a41, a42 for 4th stage
{32768, -14239, 32768}  };

int b[stages][2]=        {        //*denominator coefficients
{-11417, 25710},                  //b11, b12 for 1st stage
{-9204, 31581},                   //b21, b22 for 2nd stage
{-15860, 31605},                  //b31, b32 for 3rd stage
{-10221, 32581},                  //b41, b42 for 4th stage
{-15258, 32584}          };       //b51, b52 for 5th stage
```

FIGURE D.4. Coefficient file for an IIR bandstop filter centered at 1750 Hz, designed using MATLAB's filter designer SPTOOL (bs1750.cof).

D.3 MATLAB FOR FIR FILTER DESIGN USING THE STUDENT VERSION

FIR filters can be designed using the Student Version [2] of the MATLAB software package [1]. See also Section D.1 for the design of FIR filters using MATLAB's GUI filter designer SPTOOL.

Example D.3: FIR Filter Design Using MATLAB's Student Version

Figure D.5 shows a listing of a MATLAB program mat33.m to design a 33-coefficient FIR bandpass filter. The function remez uses the Parks–McClellan algorithm based on the Remez exchange algorithm and Chebyshev's approximation theory. The desired filter has a center frequency of 1 kHz with a sampling frequency of 10 kHz. The frequency v represents the normalized frequency variable, defined as $v = f/F_N$, where F_N is the Nyquist frequency. The bandpass filter is represented by three bands:

1. The first band (stopband) has normalized frequencies between 0 and 0.1 (0 to 500 Hz), with a corresponding magnitude of 0.
2. The second band (passband) has normalized frequencies between 0.15 and 0.25 (750 to 1250 Hz), with a corresponding magnitude of 1.
3. The third band (stopband) has normalized frequencies between 0.3 and the Nyquist frequency of 1 (1500 to 5000 Hz), with a corresponding magnitude of 0.

%Mat33.m MATLAB program for FIR Bandpass with 33 coefficients Fs=10 kHz

```
nu= [0 0.1 0.15 0.25 0.3 1];      %normalized frequencies
mag= [0 0 1 1 0 0];               %magnitude at normalized frequencies
c=remez (32,nu,mag);              %invoke remez algorithm for 33 coeff
bp33=c';                          % coeff values transposed
save matpb33.cof bp33 -ascii;     %save in ASCII file with coefficients
[h,w] =freqz (c,1,256);           %frequency response with 256 points
plot(5000*nu,mag,w/pi,abs(h))     %plot ideal magnitude response
```

FIGURE D.5. MATLAB program for FIR filter design (mat33.m).

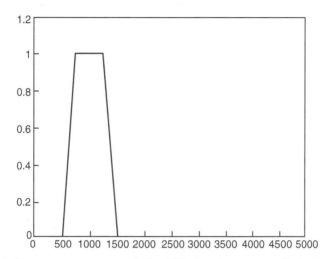

FIGURE D.6. Frequency response of the FIR bandpass filter desired, obtained with MATLAB.

Run this program from MATLAB and verify the magnitude response of the ideal desired filter plotted within MATLAB in Figure D.6. Note that the frequencies 750 and 1250 Hz represent passband frequencies with normalized frequencies of 0.15 and 0.25, respectively, and associated magnitudes of 1. The frequencies 500 and 1500 Hz represent stopband frequencies with normalized frequencies of 0.1 and 0.3, respectively, and associated magnitudes of 0. The last normalized frequency value of 1 corresponds to the Nyquist frequency of 5000 Hz and has a magnitude of zero. The program generates a set of 33 coefficients saved in the coefficient file matbp33 .cof in ASCII format.

Example D.4: Multiband FIR Filter Design Using MATLAB

This example extends the preceding three-band example to a five-band design in order to obtain two passbands. The program mat63.m (Figure D.7) is similar to the preceding MATLAB program, mat33.m. This filter with two passbands is

%Mat63.m MATLAB program for two passbands, 63 coefficients Fs=10 kHz

```
nu= [0 0.1 0.12 0.18 0.2 0.3 0.32 0.38 0.4 1]; %normalized frequencies
mag= [0 0 1 1 0 0 1 1 0 0];       %magnitude at normalized frequencies
c=remez (62,nu,mg);               %invoke remez algorithm for 63 coeff
bp63=c';                          % coeff values transposed
save mat2bp.cof bp63 -ascii;      %save in ASCII file with coefficients
[h,w] =freqz (c,1,256);           %frequency response with 256 points
plot (500*nu,mag,w/pi,abs(h))     %plot ideal magnitude response
```

FIGURE D.7. MATLAB program for a two-passband FIR filter design (mat63.m).

represented by a total of five bands: the first band (stopband) has normalized frequencies between 0 and 0.1 (0 to 500 Hz), with corresponding magnitude of 0; the second band (passband) has normalized frequencies between 0.12 and 0.18 (600 to 900 Hz), with a corresponding magnitude of 1, and so on. This is summarized as follows:

Band	Frequency (Hz)	Normalized f/F_N	Magnitude
1	0–500	0–0.1	0
2	600–900	0.12–0.18	1
3	1000–1500	0.2–0.3	0
4	1600–1900	0.32–0.38	1
5	2000–5000	0.4–1	0

Run this program from MATLAB and verify the magnitude response of the ideal two-passband filter in Figure D.8. This program generates a set of 63 coefficients saved into the coefficient file mat2bp.cof in ASCII format.

D.4 MATLAB FOR IIR FILTER DESIGN USING THE STUDENT VERSION

MATLAB can also be used for the design of IIR filters using the Student Edition of MATLAB. See also Section D.2 for the design of IIR filters using MATLAB's GUI filter designer SPTOOL.

Example D.5: IIR Filter Design Using MATLAB's Student Version

The function *yulewalk*, available in MATLAB, allows for the design of recursive filters based on a best least squares fit [1,2]. Consider again the MATLAB program mat33.m in Figure D.5 to obtain a 33-coefficient FIR bandpass filter centered at 1000 Hz. In lieu of the *remez* function for an FIR design, the MATLAB command

```
>>[a,b] = yulewalk(n,nu,mag)
```

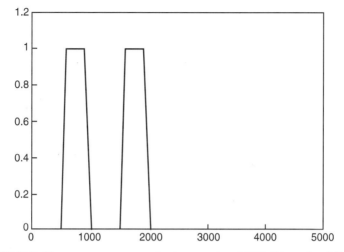

FIGURE D.8. Frequency response of a two-passband FIR filter using MATLAB.

returns the a and b coefficients in the general input–output equation in Chapter 5, associated with an IIR filter. The filter's order n represents the number of second-order sections. The C program in Example 5.1 implements an IIR filter with cascaded second-order sections, as is most commonly done. For example, if n = 6 in the `yulewalk` function, the general transfer function in Chapter 5 in terms of the resulting a and b coefficients from MATLAB needs to be reduced to one in terms of three cascaded sections.

D.5 BILINEAR TRANSFORMATION USING MATLAB AND SUPPORT PROGRAMS ON DISK

This section expands on the bilinear transformation discussion in Section 5.3.

Exercise D.1: First-Order IIR Lowpass Filter

Given a first-order lowpass analog transfer function $H(s)$, a corresponding discrete-time filter with transfer function $H(z)$ can be obtained. Let the bandwidth or cutoff frequency $B = 1$ r/s and the sampling frequency $F_s = 10\,\text{Hz}$.

1. Choose an appropriate transfer function

$$H(s) = \frac{1}{s+1}$$

which represents a lowpass filter with a bandwidth of 1 r/s.

2. Prewarp ω_D using

$$\omega_A = \tan\frac{\omega_D T}{2} = \tan\left(\frac{1}{20}\right) \cong \frac{1}{20}$$

where $\omega_D = B = 1$ r/s and $T = \frac{1}{10}$.

3. Scale $H(s)$ to obtain

$$H(s/\omega_A) = \frac{1}{20s+1}$$

4. Obtain the desired transfer function $H(z)$, or

$$H(z) = H(s/\omega_A)\big|_{s=(z-1)/(z+1)} = \frac{z+1}{21z-19}$$

Exercise D.2: First-Order IIR Highpass Filter

Given a highpass transfer function $H(s) = s/(s+1)$, obtain a corresponding transfer function $H(z)$. Let the bandwidth or cutoff frequency be 1 r/s and the sampling frequency be 5 Hz. From the preceding procedure, $H(z)$ is found to be

$$H(z) = \frac{10(z-1)}{11z-9}$$

Exercise D.3: Second-Order IIR Bandstop Filter

Given a second-order analog transfer function $H(s)$ for a bandstop filter, a corresponding discrete-time transfer function $H(z)$ can be obtained. Let the lower and upper cutoff frequencies be 950 and 1050 Hz, respectively, with a sampling frequency F_s of 5 kHz.

The transfer function selected for a bandstop filter is

$$H(s) = \frac{s^2 + \omega_r^2}{s^2 + sB + \omega_r^2}$$

where B and ω_r are the bandwidth and center frequencies, respectively. The analog frequencies are

$$\omega_{A1} = \tan\frac{\omega_{D1}T}{2} = \tan\frac{2\pi \times 950}{2 \times 5000} = 0.6796$$

$$\omega_{A2} = \tan\frac{\omega_{D2}T}{2} = \tan\frac{2\pi \times 1050}{2 \times 5000} = 0.7756$$

The bandwidth $B = \omega_{A2} - \omega_{A1} = 0.096$ and $\omega_r^2 = (\omega_{A1})(\omega_{A2}) = 0.5271$. The transfer function $H(s)$ becomes

$$H(s) = \frac{s^2 + 0.5271}{s^2 + 0.096s + 0.5271} \tag{D.1}$$

and the corresponding transfer function $H(z)$ can be obtained with $s = (z - 1)/(z + 1)$, or

$$H(z) = \frac{\{(z-1)/(z+1)\}^2 + 0.5271}{[(z-1)/(z+1)]^2 + 0.096(z-1)/(z+1) + 0.5271}$$

which can be reduced to

$$H(z) = \frac{0.9408 - 0.5827z^{-1} + 0.9408z^{-2}}{1 - 0.5827z^{-1} + 0.8817z^{-2}} \tag{D.2}$$

As shown later, $H(z)$ can be verified using the program BLT.BAS (on the accompanying disk), or MATLAB, which calculates $H(z)$ from $H(s)$ using the BLT technique, as we will illustrate. This can be quite useful in applying this procedure for higher-order filters.

Exercise D.4: Fourth-Order IIR Bandpass Filter

A fourth-order IIR bandpass filter can be obtained using the BLT procedure. Let the upper and lower cutoff frequencies be 1 and 1.5 kHz, respectively, and the sampling frequency be 10 kHz.

1. The transfer function $H(s)$ of a fourth-order Butterworth bandpass filter can be obtained from the transfer function of a second-order Butterworth lowpass filter, or

$$H(s) = H_{LP}(s)\big|_{s=(s^2+\omega_r^2)/sB}$$

where $H_{LP}(s)$ is the transfer function of a second-order Butterworth lowpass filter. $H(s)$ then becomes

$$H(s) = \frac{1}{s^2 + \sqrt{2}s + 1}\bigg|_{s=(s^2+\omega_r^2)/SB}$$

$$= \frac{s^2 B^2}{s^4 + \sqrt{2}Bs^3 + (2\omega_r^2 + B^2)s^2 + \sqrt{2}B\omega_r^2 s + \omega_r^4} \tag{D.3}$$

2. The analog frequencies ω_{A1} and ω_{A2} are

$$\omega_{A1} = \tan\frac{\omega_{D1}T}{2} = \tan\frac{2\pi \times 1050}{2 \times 10,000} = 0.3249$$

$$\omega_{A2} = \tan\frac{\omega_{D2}T}{2} = \tan\frac{2\pi \times 1500}{2 \times 10,000} = 0.5095$$

3. The center frequency ω_r and the bandwidth B can now be found, or

$$\omega_r^2 = (\omega_{A1})(\omega_{A2}) = 0.1655$$
$$B = \omega_{A2} - \omega_{A1} = 0.1846$$

4. The analog transfer function $H(s)$ is (D.3) reduces to

$$H(s) = \frac{0.03407s^2}{s^4 + 0.26106s^3 + 0.36517s^2 + 0.04322s + 0.0274} \tag{D.4}$$

5. The corresponding $H(z)$ becomes

$$H(z) = \frac{0.02008 - 0.04016z^{-2} + 0.02008z^{-4}}{1 - 2.5495z^{-1} + 3.2021z^{-2} - 2.0359z^{-3} + 0.64137z^{-4}} \tag{D.5}$$

which is in the form of (5.4). This can be verified using the program BLT.BAS (on the disk).

Exercise D.5: H(z) from H(s) Using Bilinear Function in MATLAB

Using Exercise D.3 with the second-order IIR bandstop filter, the transfer function in the analog s-plane [from (D.1)],

$$H(s) = \frac{s^2 + 0.5271}{s^2 + 0.096s + 0.5271}$$

can be converted to an equivalent transfer function in the digital z-plane using the bilinear function from MATLAB with the following commands:

```
>>num = [1, 0, 0.5271];          %numerator coefficients
>>den = [1, 0.096, 0.5271];      %denominator coefficients
>>T = 2; Fs = 1/T;               %K=1 from bilinear equation
>>[a,b]=bilinear (num, den, Fs)  %invoke bilinear function
```

to obtain the coefficients a and b associated with the transfer function in (5.4), or

$$H(z) = \frac{0.9409 - 0.5827z^{-1} + 0.9409z^{-2}}{1 - 0.5827z^{-1} + 0.8817z^{-2}}$$

which is the same transfer function (D.2) as that found in Exercise D.3. Note that $T = 2$ was chosen with MATLAB since the constant $K = 2/T$ in the bilinear equation in Chapter 5 was set to 1 for convenience. Note that MATLAB uses the following notation in the general input–output equation:

$$y(n) = b_0 x(n) + b_1 x(n-1) + b_2 x(n-2) + \cdots - a_1 y(n-1) - a_2 y(n-2) - \cdots$$

which yields a transfer function of the form

$$H(z) = \frac{b_0 + b_1 z^{-1} + b_2 z^{-2} + \cdots}{1 + a_1 z^{-1} + a_2 z^{-2} + \cdots}$$

which shows that MATLAB's a and b coefficients are the reverse of the notation used in (5.1).

Exercise D.6: Utility Program BLT.BAS to Find H(z) from H(s)

The utility program BLT.BAS (on disk), written in BASIC, converts an analog transfer function $H(s)$ into an equivalent transfer function $H(z)$ using the bilinear equation $s = (z - 1)/(z + 1)$. To verify the results in (D.1) found in Exercise D.3 for the

```
Enter the # of numerator coefficients (30 = Max, 0 = Exit) --> 3
     Enter a(0)s^2   --> 1
     Enter a(1)s^1   --> 0
     Enter a(2)s^0   --> 0.5271

Enter the # of denominator coefficients --> 3
     Enter b(0)s^2   --> 1
     Enter b(1)s^1   --> 0.096
     Enter b(2)s^0   --> 0.5271

Are the above coefficients correct ? (y/n) y
                    (a)

     a(0)z^-0  =   0.94085        b(0)z^-0  =   1.00000
     a(1)z^-1  =  -0.58271        b(1)z^-1  =  -0.58271
     a(2)z^-2  =   0.94085        b(2)z^-2  =   0.88171
                    (b)
```

FIGURE D.9. Use of BLT.BAS program for bilinear transformations: (a) coefficients in s-plane; (b) coefficients in z-plane.

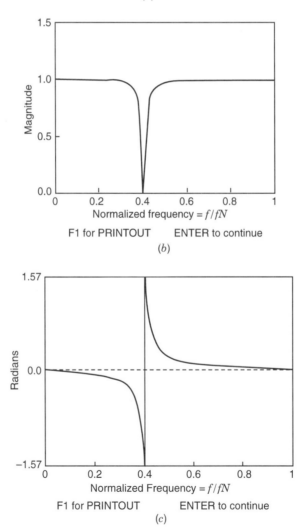

FILTER COEFFICIENTS	
NUMERATOR	DENOMINATOR
z-0 .9408	z-0 1
z-1 −.5827	z-3 −.5827
z-3 .9408	z-4 .8817
z-4	z-5
z-5	z-6
z-6	z-7
z-7	z-8
z-8	z-9
z-9	z-10
z-10	
F1 HELP F5 QUIT F10 PLOT	

(a)

F1 for PRINTOUT   ENTER to continue

(b)

F1 for PRINTOUT   ENTER to continue

(c)

FIGURE D.10. Use of the AMPLIT.CPP program for plotting magnitude and phase: (a) coefficients in the z-plane; (b) normalized magnitude; (c) normalized phase.

second-order bandstop filter, run GWBASIC, then load and run BLT.BAS. The prompts and the associated data for the a and b coefficients associated with $H(s)$ are shown in Figure D.9a and the a and b coefficients associated with the transfer function $H(z)$ are shown in Figure D.9b, which verifies (D.1). Run BLT.BAS again to verify (D.5) using the data in (D.4).

Exercise D.7: Utility Program AMPLIT.CPP to Find Magnitude and Phase

The utility program AMPLIT.CPP (on the disk), written in C++, can be used to plot the magnitude and phase responses of a filter for a given transfer function $H(z)$ with a maximum order of 10. Compile (using Borland's C++ compiler) and run this program. Enter the coefficients of the transfer function associated with the second-order IIR bandstop filter (D.2) in Exercise D.3 as shown in Figure D.10a. Figure D.10b and c show the magnitude and phase of the second-order bandstop filter. From the plot of the magnitude response of $H(z)$, the normalized center frequency is shown at $v = f/F_N = 1000/2500 = 0.4$.

Run this program again to plot the magnitude response associated with the fourth-order IIR bandpass filter in Exercise D.4. Verify the plot shown in Figure D.11. The normalized center frequency is shown at $v = 1250/5000 = 0.25$.

A utility program MAGPHSE.BAS (on the disk), written in BASIC, can be used to tabulate the magnitude and phase responses.

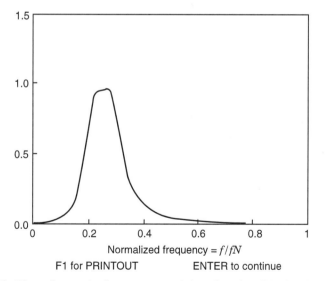

Normalized frequency = f/fN

F1 for PRINTOUT   ENTER to continue

FIGURE D.11. Plot of magnitude response of fourth-order IIR bandpass filter using AMPLIT.CPP.

D.6 FFT AND IFFT

MATLAB can be used to find both the fast Fourier transform FFT of a sequence of numbers and the inverse Fourier transform IFFT.

Exercise D.8: Eight-Point FFT and IFFT Using MATLAB

The eight-point FFT in Exercise 6.1 can readily be verified with MATLAB, with the following commands:

```
>>x = [1 1 1 1 0 0 0 0];
>>y = fft(x)
>>magy = abs(y)
>>plot (magy)
```

The resulting output magnitude transform is also plotted.

Similarly, the inverse FFT can also be verified. Given the output sequence X's in Exercise 6.1, the inverse FFT or IFFT can be found:

```
>>X = [4 1-2.414*i 0 1-0.414*i 0 1+0.414*i 0 1+2.414*i];
>>y = ifft(X)
```

where y is the resulting rectangular sequence.

REFERENCES

1. *MATLAB, The Language of Technical Computing*, MathWorks, Natick, MA 2000.

2. *MATLAB Student Version*, MathWorks, Natick, MA.

3. W. J. Gomes III and R. Chassaing, Filter design and implementation using the TMS320C6x interfaced with MATLAB, *Proceedings of the 1999 ASEE Annual Conference*, 1999.

4. W. J. Gomes III and R. Chassaing, Real-time FIR and IIR filter design using MATLAB interfaced with the TMS320C31 DSK, *Proceedings of the 1999 ASEE Annual Conference*, 1999.

5. R. Chassaing, *Digital Signal Processing Laboratory Experiments Using C and the TMS320C31 DSK*, Wiley, New York, 1999.

E

Additional Support Tools

The following additional support tools are available (see also Appendix D for MATLAB support):

1. Goldwave utility for signal generation, virtual instrument, etc.
2. FIR and IIR filter design using digifilter from MultiDSP
3. Homemade filter development package
4. Visual Application Builder (VAB)
5. Codec support from Integrated-DSP
6. Developer's kit from MATLAB

E.1 GOLDWAVE SHAREWARE UTILITY AS VIRTUAL INSTRUMENT

Goldwave is a shareware utility software program that can turn a PC with a sound card into a virtual instrument. It can be downloaded from the Web [1]. One can create a function generator to generate different signals such as sine wave and random noise. It can also be used as an oscilloscope, as a spectrum analyzer, and to record/edit a speech signal. Effects such as echo and filtering can be obtained. Lowpass, highpass, bandpass, and bandstop filters can be implemented on a sound card with Goldwave and their effects on a signal illustrated readily.

Goldwave was used to obtain an input voice (*TheForce.wav*, on the disk) added with two sinusoidal signals of frequencies 900 and 2700 Hz, respectively. This corrupted voice signal, shown in Figure 4.24, is used in Example 4.7 to illustrate removal of the two sinusoidal signals.

One can use two copies of Goldwave running under Windows 9x: one to generate a signal as input to the DSK, another to use the DSK's output into the sound card as a spectrum analyzer. However, the results obtained running two copies of Goldwave can be quite noisy.

Other shareware utility programs, such as Cool Edit [2] or Spectrogram [3], also can be used as virtual spectrum analyzers.

E.2 FILTER DESIGN USING DIGIFILTER

DigiFilter is a filter design package for the design of both FIR and IIR filters [4]. Currently, it interfaces to the C31 DSK for real-time implementation.

E.2.1 FIR Filter Design

Figure E.1 shows a plot of the log magnitude response of a 61-coefficient FIR bandpass filter centered at 2 kHz using the Kaiser window function. For a specific design, the user can select among several window functions, with the specification of the number of taps (coefficients) associated with each window (rectangular, Hamming,

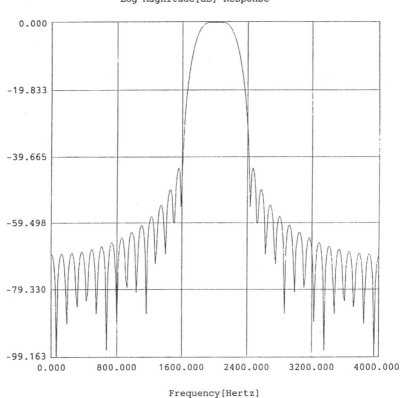

FIGURE E.1. Magnitude response of FIR bandpass filter using DigiFilter.

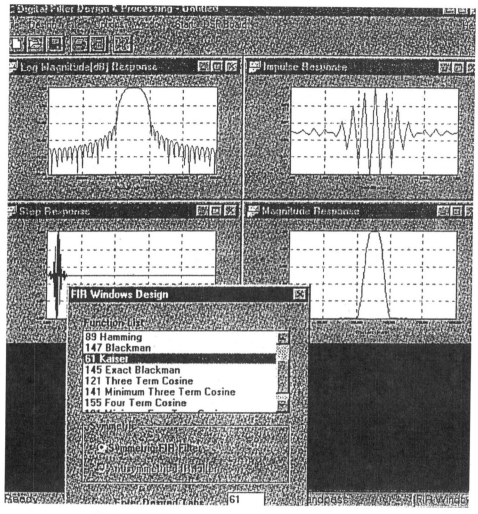

FIGURE E.2. Responses of FIR filter using DigiFilter.

etc.). Impulse as well as step responses can also be obtained, as shown in Figure E.2. Note that an implementation with a Hamming window function would require 89 coefficients, whereas a Kaiser window would require 61 coefficients (Figure E.2).

E.2.2 IIR Filter Design

An IIR filter can readily be designed with the filter package DigiFilter. One can choose among several designs using the following functions: Butterworth, Chebyshev, elliptic, and Bessel, each associated with a specific filter order. A plot of the magnitude response similar to an FIR design, as well as a plot of the poles and zeros of $H(z)$, can be obtained.

E.3 FIR FILTER DESIGN USING FILTER DEVELOPMENT PACKAGE

A noncommercial filter development package appears on the accompanying disk. The program `FIRprog.bas`, written in BASIC, calculates the coefficients of an FIR filter. This program is discussed in Refs. 5 to 7. It allows for the design of low-pass, highpass, bandpass, and bandstop FIR filters using the rectangular, Hanning, Hamming, Blackman, and Kaiser window functions. The resulting coefficients can be generated in integer or float format. This file needs to be modified and incorporated into one of the generic FIR programs.

E.3.1 Kaiser Window

1. Run BASIC and load/run the program `FIRprog.bas`. Figure E.3*a* and *b* show a display of available window functions and the frequency-selective filters that can be designed. Select the Kaiser window option and a bandpass filter. A separate module for the Kaiser window (`FIRproga.bas`) is called from `FIRprog.bas`.

2. Enter the specifications shown in Figure E.3*c*. Choose the c31 option to save the 53 resulting coefficients into a file in a float format (the C25 option saves the coefficients in hexadecimal). Save it as `BP53K.cof`.

3. Edit it (an edited version is on the disk). Include it in the program `FIRPRN.c` in Example 4.4. Build/run and verify the frequency response of the FIR bandpass filter centered at 800 Hz shown in Figure E.4, obtained with an HP analyzer. An internally generated noise sequence becomes the input to the FIR filter in the program `FIRPRN.c`. This filter was designed so that the center frequency is at 1000 Hz ($F_s/10$), selecting a sampling frequency of 10,000 Hz. Since we are using a sampling frequency of 8 kHz with the DSK, the center frequency is at 800 Hz, as shown in Figure E.4.

E.3.2 Hamming Window

Repeat this procedure for a Hamming window function. Enter 900 and 1100 for the lower and upper cutoff frequencies. Enter 5.2 (ms) for the duration D of the impulse response, since the number of coefficients N is

$$N = (D \times F_s) + 1$$

This will yield a design with 53 coefficients. Save the resulting coefficient file as `BP53H.cof`. Edit it as with the Kaiser window, test it using the program `FIRPRN.c`, and verify an FIR bandpass filter with a narrower mainlobe.

E.4 VISUAL APPLICATION BUILDER

The Visual Application Builder (VAB), available from Hyperception [8], is a component-based virtual design tool that can be used to implement DSP algorithms.

```
                        Main Menu
                        _____


                        1. . . .RECTANGULAR
                        2. . . .HANNING
                        3. . . .HAMMING
                        4. . . .BLACKMAN
                        5. . . .KAISER
                        6. . . .Exit to DOS

        Enter window desired (number only) -> 5
                           (a)

               Selections:
                        1. . . .LOWPASS
                        2. . . .HIGHPASS
                        3. . . .BANDPASS
                        4. . . .BANDSTOP
                        5. . . .Exit back to Main Menu

      Enter desired filter type (number only) -> 3
                           (b)

      Specifications:
                 BANDPASS
                 Passband Ripple (AP) = 6 db
                 Stopband Attenuation (AS) = 30 db
                 Lower Passband Frequency = 900 Hz
                 Upper Passband Frequency = 1100 Hz
                 Lower Stopband Frequency = 600 Hz
                 Upper Stopband Frequency = 1400 Hz
                 Sampling Frequency (Fs) = 10000 Hz

      The calculated # of coefficients required is: 53

      Enter # of coefficients desired ONLY if greater than 53
      otherwise, press <Enter> to continue ->
                           (c)

Send coefficients to:
                 (S)creen
                 (P)rinter
                 (F)ile: contains TMS320 (C25 or C31) data format
                 (R)eturn to Filter Type Menu
                 (E)xit to DOS

Enter desired path --> f

Enter DSP type (C25 OR C31):? c31
                           (d)
```

FIGURE E.3. FIR filter design with filter development package (on disk): (*a*) choice of windows: (*b*) type of filter; (*c*) filter specifications; (*d*) menu for coefficients format.

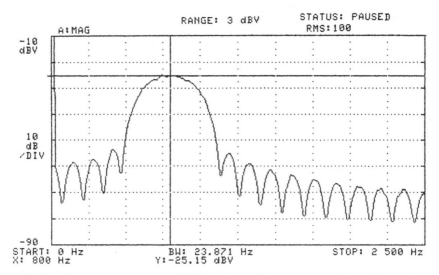

FIGURE E.4. Frequency response of FIR bandpass filter using coefficient file `BP53K.cof` generated with filter package on disk.

VAB uses a methodology of developing DSP algorithms and systems graphically simply by connecting functional components together with a mouse. The user only needs to choose the desired functions, place them onto a worksheet, select their parameters interactively, and describe the data flow using line connections. The method of design is quite similar to drawing a block diagram of the system being designed. DSP-based design implementations can be created and executed on DSP hardware without having to write any source code at all.

VAB contains a wide range of functional block components for FFT, filtering, and so on, and supports the C6711 DSK. Within a few minutes, one can design and test a DSP system that includes functional blocks such as signal generators, A/D and D/A, filters, FFT, image processing components, and so on. Results can be quickly displayed on the PC monitor as the algorithm is executing or to an external device such as an oscilloscope. Figure E.5 shows a block diagram of a Vocoder implemented on the C6711 DSK.

E.5 MISCELLANEOUS SUPPORT

The following additional support tools are available (see also Appendix D on MATLAB and Appendix F on the Audio Daughter Card based on the PCM3003 codec):

1. Daughter card based on the AD77 stereo codec that interfaces to the C6x DSK, available from Integrated-DSP [9].
2. Developer's kit for Texas Instruments' DSP [10], which connects MATLAB and SIMULINK with Texas Instruments' software and hardware. It focuses

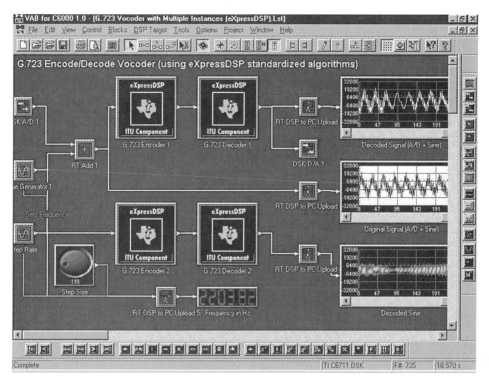

FIGURE E.5. Vocoder block diagram implemented on the C6711 DSK using VAB.

on code optimization and test and analysis rather than rewriting DSP algorithms. It currently supports the C6701-based evaluation module (EVM).

REFERENCES

1. Goldwave, at *www.goldwave.com.*

2. Cool Edit, at *www.syntrillium.com.*

3. Gram412.zip from Spectrogram, address from shareware utility with the database address *www.simtel.net.*

4. DigiFilter, from MultiDSP, at *multidsp@aol.com.*

5. R. Chassaing, *Digital Signal Processing Laboratory Experiments Using C and the TMS320C31 DSK,* Wiley, New York, 1999.

6. R. Chassaing, *Digital Signal Processing with C and the TMS320C30,* Wiley, New York, 1992.

7. R. Chassaing and D. W. Horning, *Digital Signal Processing with the TMS320C25,* Wiley, New York, 1990.

8. Hyperception, at *info@hyperception.com.*

9. Integrated DSP, at *www.integrated-dsp.com.*

10. The MathWorks, Inc., at *www.mathworks.com.*

F

Input and Output with PCM3003 Stereo Codec

F.1 PCM3003 AUDIO DAUGHTER CARD

The PCM3003 stereo codec [1,2] provides an alternative to the AD535 codec. It has a higher sampling rate, up to approximately 73 kHz, and two complete input and output channels. A different communication program, *C6xdskinit_pcm.c*, is used with the PCM3003 (in lieu of *C6xdskinit.c*), and also a different header file, *C6xdskinit_pcm.h*, which contains the functions prototypes (in lieu of *C6xdskinit.h*). Several examples are included to illustrate the use of the PCM3003 stereo codec with two inputs.

Figure F.1 shows a schematic diagram of an inexpensive ($50) PCM3003 audio daughter card, available from TI, that can be plugged to the DSK. It can also be interfaced with the TMS320C3x. It plugs into an 80-pin connector JP3 on the DSK (another 80-pin connector J1 on the DSK contains data and address lines).

A jumper can be set through connector JP5 on the audio daughter card for either a fixed sampling frequency of 48 kHz (desired and actual) or for a programmable desired sampling rate. From Figure F.1, with a jumper in position 3–4, a fixed sampling rate of 48 kHz can be obtained, since this connects to a 12.288-MHz clock on board the audio card, yielding

$$F_s = 12.288 \, \text{MHz}/256 = 48 \, \text{kHz}$$

With the jumper in position 1–2 a variable F_s can be obtained using timer 0. A desired sampling rate F_s (unless fixed at 48 kHz) can be specified/set in the program. F_s is global and the actual sampling rate is calculated within the communication support file *C6xdskinit_pcm.c*. The following illustrates some desired sampling frequencies and corresponding actual sampling frequencies:

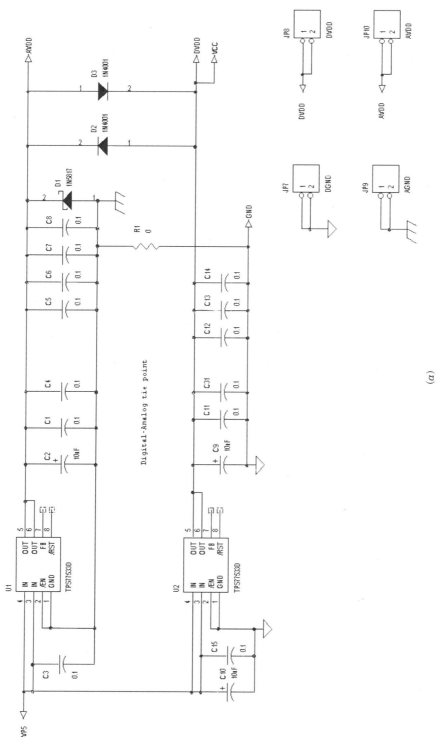

FIGURE F1. Schematic of PCM3003-based audio daughter card that interfaces to C6711 DSK (Courtesy of Texas Instruments).

(a)

311

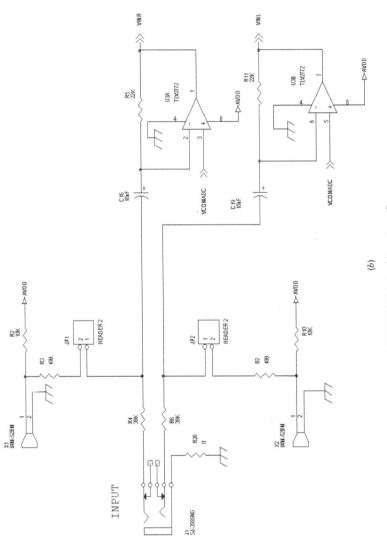

FIGURE F.1. (*Continued*)

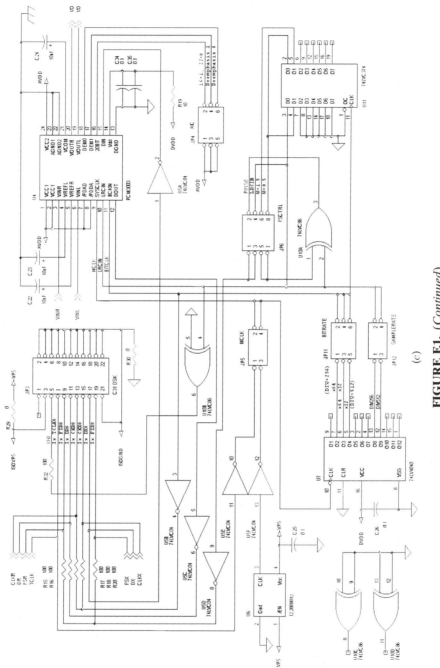

FIGURE F.1. (*Continued*)

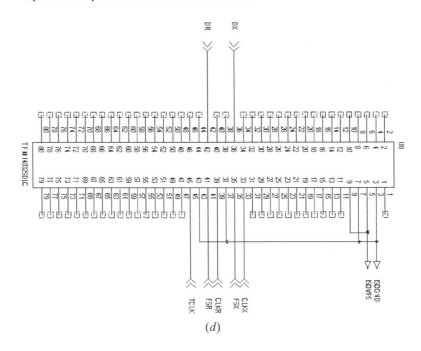

(*d*)

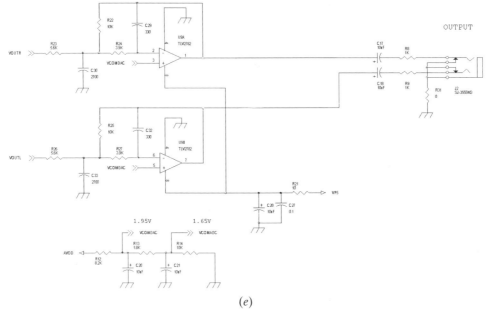

(*e*)

FIGURE F.1. (*Continued*)

F_s Desired (Hz)	Actual F_s (Hz)	
8,000	8,138.021	
16,000	14,648.438	
20,000	18,310.547	
48,000	36,621.094	(jumper position in JP5 for variable rate)
48,000	48,000	(jumper position in JP5 for fixed rate)
>48,000	73,242.187	(jumper position in JP5 for variable rate)

For a variable sampling rate, F_s is calculated within the program $C6xdskinit\_$ $pcm.c$ using a desired frequency (set in the program), a clock frequency of 150 MHz/ 4, and clocks per sample as 256. A maximum sampling rate of 73,242.18 Hz can be obtained (though F_s > 48,000 is not recommended by TI).

Two dedicated connectors (stereo to mono) are used for the examples in this appendix. This type of connector has two input and one single-ended output connections. A 16-bit data value is obtained from each input channel, and the resulting single-ended output connection yields 32-bit data (16 bits from each channel). This output connection with 32-bit data connects to the input PCM3003 codec. The two inputs connections are designated by silver for the left channel and gold for the right channel.

F.2 PROGRAMMING EXAMPLES USING THE PCM3003 STEREO CODEC

Example F.1: Loop Program Using Polling with the PCM3003 Stereo Codec (loop_poll_pcm)

Figure F.2 shows a listing of the program $loop\_poll\_pcm.c$, which implements a loop using the PCM3003 codec. See also Example 2.2, which implements a loop using the onboard AD535 codec.

Variable F_s

A desired frequency of F_s = 16,000 Hz is specified in the program. The jumper in JP5 should be in position 1–2. The actual sampling frequency is calculated within $C6xdskinit\_pcm.c$ as

$$F_s(actual)=14,648.438 Hz$$

with a divider value of 5 (divider cast as integer).

Build this project as **loop_poll_pcm.** Include the two source files $C6xdskinit\_pcm.c$ and $vectors.asm$, along with $loop\_poll\_pcm.c$.

Input a sinusoidal signal with an amplitude of approximately 1 V and a frequency of 1 kHz. Observe the corresponding output as the delayed input. Increase the

//loop_poll_pcm.c Loop program with polling using PCM3003 codec

```
float Fs = 16000.0;                          //desired (Actual=14,648 Hz)

void main()
{
  comm_poll();                               //init DSK,codec,McBSP
  while(1)                                    //infinite loop
    output_left_sample(input_left_sample()); //IN from left,OUT from left
}
```

FIGURE F.2. Loop program with polling using PCM3003 codec (loop_poll_pcm.c).

frequency beyond 7 kHz. Verify that the bandwidth of the antialiasing filter is approximately 6.8 kHz.

Select View → Quick Watch window to watch *Fs_actual*, and verify that it is calculated (displayed) as 14,648.438 Hz.

Fixed *F_s* = 48 kHz

Set the jumper in JP5 to position 3–4 for a fixed sampling rate. Setting F_s in the program is irrelevant. Rebuild and run. Figure F.3 shows the output of the codec displayed on an HP analyzer using noise as input. It illustrates that the bandwidth of the antialiasing filter is approximately 21.5 kHz.

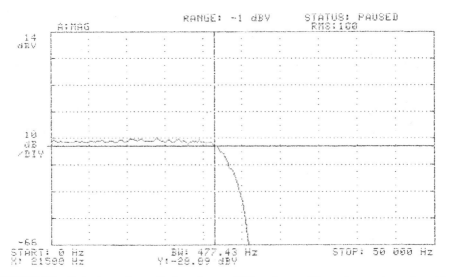

FIGURE F.3. Output spectrum displayed on an HP analyzer with random noise as input for a fixed sampling rate of $F_s = 48$ kHz (using loop_poll_pcm).

```
//loop_intr_pcm.c Loop program with interrupt using PCM3003

float Fs = 16000.0;              //irrelevant since jumper in 3-4

interrupt void c_int11()        //interrupt service routine
{
      output_left_sample(input_left_sample()); //IN/OUT from left

      return;                   //return from interrupt
}

void main()
{
      comm_intr();              //init DSK, codec, McBSP
      while(1);                 //infinite loop
}
```

FIGURE F.4. Loop program with interrupt using a PCM3003 codec (`loop_intr _pcm.c`).

Increase the amplitude of a sinusoidal signal as input to verify that the output saturates beyond an input voltage of approximately 3.5 V p-p.

Experiment with input and output from different channels. For example,

output_sample(input_sample());

acquires a 32-bit data item (16 bits from each channel). A mono connector can be used and defaults to the left channel. However,

output_right_sample(input_left_sample);

requires that the stereo-to-mono connector obtain an output from the right channel (gold) with an input from the left channel (silver).

Example F.2: Loop Program Using Interrupt with the PCM3003 Codec (loop_intr_pcm)

This example illustrates an interrupt-driven version of the loop program using the PCM3003 codec. Example F.1 illustrates the loop feature using polling. See also Example 2.1, use of the onboard AD535 codec. Figure F.4 shows a listing of *loop_intr_pcm.c* that implements this example.

`//`**Fir_pcm.c** FIR using PCM3003 codec

```
#include "bp41.cof"                //coefficient file BP @ Fs/8
int yn = 0;                        //initialize filter's output
short dly[N];                      //delay samples
float Fs = 48000.0;                //fixed/actual Fs

interrupt void c_int11()           //ISR
{
  short i;

  dly[0] = input_left_sample();    //newest input @ top of buffer
  yn = 0;                          //initialize filter's output
  for (i = 0; i< N; i++)
          yn += (h[i] * dly[i]);   //y(n)+=h(i)*x(n-i)
  for (i = N-1; i > 0; i--)        //starting @ bottom of buffer
          dly[i] = dly[i-1];       //update delays with data move

  output_right_sample(yn >> 15);   //output filter
  return;                          //return from ISR
}

void main()
{
  comm_intr();                     //init DSK, codec, McBSP
  while(1);                        //infinite loop
}
```

FIGURE F.5. FIR program using a PCM3003 codec (`FIR_pcm.c`).

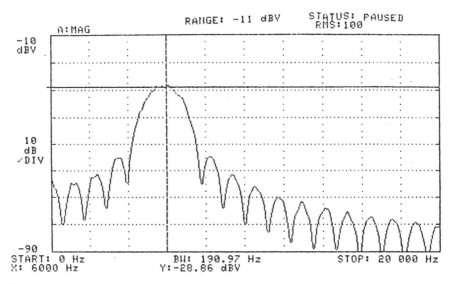

FIGURE F.6. Output frequency response of an FIR bandpass filter centered at $F_s/8$ obtained with an HP analyzer.

Build this project as **loop_intr_pcm**. Verify similar results as with the polling version in Example F.1, with F_s fixed at 48 kHz (jumper in position 3–4).

Example F.3: FIR Filter Implementation Using the PCM3003 Codec (FIR_pcm)

Figure F.5 shows a listing of the program FIR_pcm.c, which implements an FIR filter using the PCM3003 codec. Example 4.4 illustrates the implementation of an FIR filter using the onboard codec AD535. The filter coefficient bp41.cof represents a 41-coefficient FIR bandpass filter centered at $F_s/8$ (used in Chapter 4). The sampling frequency is set and fixed at 48 kHz (using jumper JP5 in position 3–4).

Build this project as **FIR_pcm**. Figure F.6 shows the frequency response of the FIR filter using noise as input, obtained with an HP analyzer. An actual (using the jumper position 3–4 for fixed rate) sampling frequency of 48 kHz is used. The center frequency is shown as 6 kHz, corresponding to $F_s/8$.

Change the jumper for a variable sample rate (position 1–2) and set F_s to 60 kHz in the program (or set to any frequency greater than 48 kHz and up to 73 kHz). The variable divider, calculated in C6xdskinit_pcm.c, is 1 for this range of frequencies. Rebuild/run this project and verify a band pass filter centered at 73, 248/8 = 9.15 kHz.

Example F.4: Adaptive FIR Filter for Noise Cancellation Using the PCM3003 Codec (adaptnoise_pcm)

Figure F.7 shows a listing of the program Adaptnoise_pcm.c, which illustrates the noise canceler using the PCM3003 stereo codec. See also Example 7.2, which implements the noise canceler using the onboard AD535 codec. The desired sampling frequency is set at 8 kHz in the program; but the actual rate is 8138.021 Hz. Build this project as **adaptnoise_pcm.**

1. *Desired: 1.5 kHz, undesired: 2 kHz.* Input a desired sinusoidal signal (with a frequency such as 1.5 kHz) into the left channel and an undesired sinusoidal noise signal of 2 kHz into the right channel. Run the program. Verify that the 2-kHz noise signal is being canceled gradually (you can adjust the rate of convergence by changing *beta* by a factor of 10 in the program). Access the slider gel program adaptnoise.gel and change the slider to position 2. Verify the output as the two original sinusoidal signals at 1.5 and at 2 kHz.

2. *Desired: wideband random noise; undesired: 2 kHz.* Input random noise (from Goldwave or noise generator) as the *desired* wideband signal into the left channel, with the *undesired* 2-kHz sinusoidal signal into the right input

```
//Adaptnoise_pcm.c   Adaptive FIR for noise cancellation using PCM3003

#define beta 1E-10                       //rate of convergence
#define N 30                             //# of weights (coefficients)
#define LEFT 0                           //left channel
#define RIGHT 1                          //right channel
float w[N];                              //weights for adapt filter
float delay[N];                          //input buffer to adapt filter
float Fs = 8000.0;                       //sampling rate
short output;                            //overall output
short out_type = 1;                      //output type for slider
volatile union{unsigned int uint; short channel[2];}CODECData;

interrupt void c_int11()                 //ISR
 {
  short i;
  float yn=0, E=0, dplusn=0, desired=0, noise=0;

  CODECData.uint = input_sample();       //input 32-bit from both channels
  desired = (float) CODECData.channel[LEFT];  //input left channel
  noise = (float) CODECData.channel[RIGHT];   //input right channel

  dplusn = desired + noise;              //desired+noise
  delay[0] = noise;                      //noise as input to adapt FIR

  for (i = 0; i < N; i++)                //to calculate out of adapt FIR
      yn += (w[i] * delay[i]);           //output of adaptive filter

  E = (desired + noise) - yn;            //"error" signal=(d+n)-yn

  for (i = N-1; i >= 0; i--)             //to update weights and delays
   {
      w[i] = w[i] + beta*E*delay[i];     //update weights
      delay[i] = delay[i-1];             //update delay samples
   }
  if (out_type == 1)                     //if slider in position 1
      output = ((short)E);               //error signal as overall output
  else if (out_type == 2)
      output=((short)dplusn);            //desired+noise
  output_left_sample(output);            //overall output result
  return;
}

 void main()
{
  short T=0;
  for (T = 0; T < 30; T++)
   {
      w[T] = 0;                          //init buffer for weights
      delay[T] = 0;                      //init buffer for delay samples
   }
  comm_intr();                           //init DSK, codec, McBSP
  while(1);                              //infinite loop
}
```

FIGURE F.7. Program that implements adaptive noise canceler using the PCM3003 codec (adaptnoise_pcm.c).

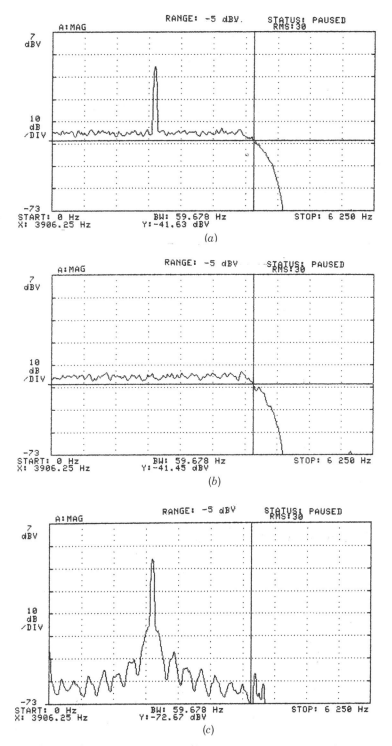

FIGURE F.8. Output frequency responses (from `adaptnoise_pcm.c`) displayed on an HP analyzer: (*a*) desired wideband random signal and undesired 2-kHz sinusoidal signal; (*b*) desired wideband random signal with undesired 2-kHz signal canceled; (*c*) desired 2-kHz signal with wideband random signal canceled.

```
//Adaptpredict_pcm.c Adaptive predictor to cancel interference
#define beta 1E-15              //rate of convergence
#define N 60                    //# of coefficients of adapt FIR
#define NS 256                  //size of wideband's buffer
#define LEFT 0                  //left channel
#define RIGHT 1                 //right channel
const short bufferlength = NS;  //buffer length for wideband signal
short splusn[N+1];              //buffer wideband signal+interference
float w[N+1];                   //buffer for weights of adapt FIR
float delay[N+1];               //buffer for input to adapt FIR
float Fs = 48000.0;             //for fixed Fs
volatile union {unsigned int uint; short channel[2];}CODECData;

interrupt void c_int11()        //ISR
{
 static short buffercount=0;    //init buffer
 short i;
 float yn, E;                   //yn=out adapt FIR, error signal
 short wb_signal;               //wideband desired signal
 short noise;                   //external interference

 CODECData.uint = input_sample(); //input left and right as 32-bit
 wb_signal = (float) CODECData.channel[LEFT]; //desired on left channel
 noise = (float) CODECData.channel[RIGHT];    //noise on right channel

 splusn[0] = (wb_signal + noise); //wideband signal+interference
 delay[0] = splusn[3];          //delayed input to adaptive FIR
 yn = 0;                        //init output of adaptive FIR

 for (i = 0; i < N; i++)
   yn += (w[i] * delay[i]);     //output of adaptive FIR filter

 E = splusn[0] - yn;            //(wideband+noise)-out adapt FIR

 for (i = N-1; i >= 0; i--)
   {
   w[i] = w[i]+(beta*E*delay[i]); //update weights of adapt FIR
   delay[i+1] = delay[i];       //update buffer delay samples
   splusn[i+1] = splusn[i];     //update buffer corrupted wideband
   }

 buffercount++;                 //incr buffer count of wideband
 if (buffercount >= bufferlength) //if buffer count=length of buffer
   buffercount = 0;             //reinit count
 output_left_sample((short)E);  //overall output from left channel
 return;
}

void main()
{
 int T = 0;
 for (T = 0; T < N; T++)        //init variables
   {
   w[T] = 0.0;                  //init weights of adaptive FIR
   delay[T] = 0.0;              //init buffer for delay samples
   splusn[T] = 0;               //init wideband+interference
   }
 comm_intr();                   //init DSK, codec, McBSP
 while(1);                      //infinite loop
}
```

FIGURE F.9. Adaptive predictor program using a PCM3003 codec (adaptpredict_
pcm.c).

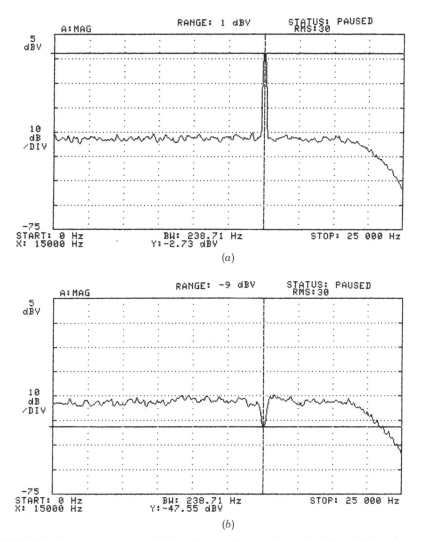

FIGURE F.10. Output spectrum of adaptive predictor obtained with an HP analyzer; (*a*) desired wideband random signal and 15-kHz narrowband interference; (*b*) desired wideband random signal with 15-kHz interference canceled.

channel. Restart/run the program. Access the slider and change it to position 2. Figure F.8*a* shows the output spectrum of both the desired wideband signal and the additive undesired 2-kHz sinusoidal signal, obtained with an HP analyzer (with the slider in position 2). Figure F.8*b* shows the undesired 2-kHz signal canceled, displaying the wideband signal as the output spectrum (with the slider in position 1). Verify the gradual cancellation of the undesired 2-kHz signal.

3. *Desired: 2 kHz; undesired: wideband random noise.* Switch the inputs to the connector so that the desired 2-kHz signal is the left-channel input and the undesired wideband random noise signal is the right-channel input. Increase beta by 100. Rebuild/run the program. Verify the gradual cancellation of the undesired random noise signal (with the slider in position 1). Figure F.8c shows the 2-kHz signal with the undesired wideband noise signal canceled out.

Example F.5: Adaptive Predictor for Cancellation of Narrowband Interference Added to Desired Wideband Signal, Using the PCM3003 Codec (adaptpredict_pcm)

Figure F.9 shows a listing of the program *adaptpredict_pcm* for the cancellation of a narrowband interference in the presence of a wideband signal. This example uses the PCM3003 codec. See also Example 7.6, which implements the adaptive predictor using the onboard AD535 codec. A sampling rate of 48 kHz (desired/ actual) is used with the jumper JP5 for a fixed sample rate position.

Build this project as **adaptpredict_pcm.** Input random noise as the desired wideband random signal (from Goldwave, noise generator, etc.), and a 15-kHz signal as an undesired narrowband interference. Figure F.10a shows the output spectrum of the wideband random signal with the 15-kHz additive narrowband interference. Figure F.10b shows the narrowband additive interference canceled. Verify the gradual cancellation of the 15-kHz interference.

REFERENCES

1. *PCM3002/PCM3003 16-/20-Bit Single-Ended Analog Input/Output Stereo Audio Codec*, SBAS079, Burr-Brown/Texas Instruments, Dallas, TX, 2000.

2. *TMS320C6000 McBSP: I²S Interface*, SPRA595, Texas Instruments, Dallas, TX, 1999.

G

DSP/BIOS and RTDX for Real-Time Data Transfer

DSP/BIOS provides CCS the capability for analysis, scheduling, and data exchange in real time [1–5]. An application program can be analyzed while the digital signal processor is running (the target processor need not be stopped). There are many DSP/BIOS application programming interface (API) modules available for real-time analysis, input/output, and so on. API functions are included with CCS to configure and control operation of the codec. They initialize the DSK, the McBSP, and the codec.

1. *Real-time analysis.* This can be either critical or not so critical. For example, one needs to respond to input samples so that information is not lost. On the other hand, the transfer of data from the digital signal processor to the host PC may be done between incoming samples.

2. *Real-time scheduling.* Data transfer is scheduled through DSP/BIOS software interrupts. Tasks/functions are initially assigned different priorities. Based on results obtained from a CPU execution graph, one can reprioritize these different tasks. The CPU execution graph shows when various tasks are executed, and whether or not the CPU misses real-time data. This graph is similar to the type of plots obtained with a logic analyzer. An execution graph associated with an audio example (included with CCS) is shown in Figure G.1. This graph shows the execution of *threads*. A thread can be an independent stream of instructions executed by the DSP processor. It may contain an ISR, a function call, and so on. Different types of threads are given different priorities. Hardware interrupts (HWIs) have the highest priorities, followed by software interrupts (SWI), which include periodic functions (PRD).

3. *Real-time data exchange* (RTDX). This allows the exchange of data between the host and the processor, via the Joint Test Action Group (JTAG) interface,

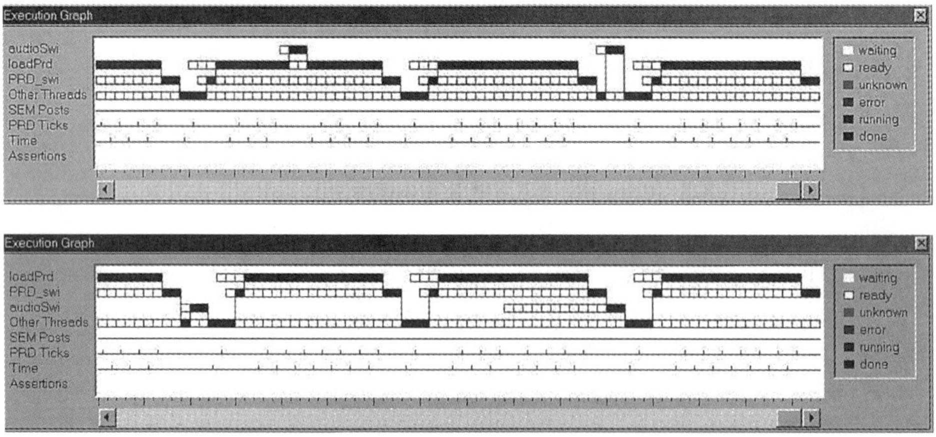

(b)

FIGURE G.1. CCS plot of execution graphs as CPU is being overloaded with NOPs: (a) output not degraded when setting audioSwi with the highest priority; (b) output degraded when setting audioSwi with lower priority.

while the processor is running. RTDX consists of both target and host components. Data are transferred through two "pipes" (one for receiving and one for transmitting). If the CPU starts missing real-time data, one can find out from the execution graph. Reprioritizing, if possible, could then solve this problem.

Examples of DSP/BIOS with RTDX

An audio example is included with the DSK package. It is essentially a "loop" example. It can illustrate overloading the CPU. This is accomplished by executing NOPs. As the number of NOPs is increased, the effects on the output can be monitored. Figure G.1a indicates that the task of "audioSwi" has the highest priority and can interrupt the lower priority task of "loadPrd." In Figure G.1b, "audioSwi" has a lower priority and has to wait for the higher-priority tasks of loadPrd and Prd_swi. This causes data to be missed. For example, with music as input, and the number of NOPs increasing (up to a million), one can hear the gradual degradation of the output signal as the CPU starts missing execution. The execution graph can show when the CPU starts missing data.

Another example included with CCS makes use of the LOG module LOG_printf() to monitor a program in real time. The C function *printf()*, supported by real-time library support, takes too many cycles to be desirable for real-time monitoring; the LOG module LOG_printf() takes considerably less time. The LOG_printf() function can be used to record data in critical time while the transfer of data from the target processor to the host can occur in not so critical time. Results on the performance of LOG_printf() supported with DSP/BIOS versus `printf()` supported with the runtime support library show that `printf()` can take 100 times more cycles to execute.

The project example PLL, discussed in Chapter 9, includes the code version (on the disk) associated with DSP/BIOS's RTDX.

REFERENCES

1. *TMS320C6000 DSP/BIOS User's Guide,* SPRU303B, Texas Instruments, Dallas, TX, 2000.

2. *An Audio Example Using DSP/BIOS,* SPRA598, Texas Instruments, Dallas, TX, 1999.

3. *TMS320C6000 DSP/BIOS Application Programming Interface (API) Reference Guide,* SPRU403A, Texas Instruments, Dallas, TX, 2000.

4. *Application Report, DSP/BIOS by Degrees: Using DSP/BIOS Features in an Existing Application,* SPRA591, Texas Instruments, Dallas, TX, 1999.

5. *Real-Time Data Exchange,* SPRY012, Texas Instruments, Dallas, TX, 1998.

Index